# Electrical Systems for Architects

## Aly S. Dadras, N.C.A.R.B.

*Architect - Planner*

*Professor of Architecture*

*New York Institute of Technology*

**McGraw-Hill, Inc.**

New York  San Francisco  Washington, D.C.  Auckland  Bogotá
Caracas  Lisbon  London  Madrid  Mexico City  Milan
Montreal  New Delhi  San Juan  Singapore
Sydney  Tokyo  Toronto

**Library of Congress Cataloging-in-Publication Data**

Dadras, Aly S.
    Electrical systems for architects / Aly S. Dadras.
        p.    cm.
    Includes bibliographical references.
    ISBN 0-07-015078-8 (alk. paper)
    1. Buildings—Electric equipment—Handbooks, manuals, etc.
2. Electric engineering—Handbooks, manuals, etc.  3. Buildings—
Mechanical equipment—Handbooks, manuals, etc.  I. Title.
TK4035.B83D34   1995
621.31—dc20                       94-47176
                                    CIP

1 2 3 4 5 6 7 8 9 0  KGP/KGP  9 0 0 9 8 7 6 5

ISBN 0-07-015078-8

*The sponsoring editor for this book was Joel Stein, the editing supervisor was Christine Furry, and the production supervisor was Suzanne W. B. Rapcavage. It was set in Helvetica by North Market Street Graphics.*

*Printed and bound by Arcata Graphics/Kingsport Press.*

McGraw-Hill books are available at special quantity discounts to use as premiums and sales promotions, or for use in corporate training programs. For more information, please write to the Director of Special Sales, McGraw-Hill, Inc., 11 West 19th Street, New York, NY 10011. Or contact your local bookstore.

This book is printed on acid-free paper.

*Dedicated*

*to the*

**New York Institute of Technology**

*where I have been honored and privileged to be a part of its*

*distinguished faculty for thirty years,*

*to the*

***University of Miami***

*where I received my undergraduate education and was*

*honored to receive the first*

*"Alumnus of Distinction Award" in 1970,*

*and to*

***Every one of my thousands of students***

*who have helped me to learn and*

*encouraged me to write.*

# About the Author

**Aly S. Dadras** is a registered architect and professor of architecture at the New York Institute of Technology, where he was one of the founders of the School of Architecture in 1964. He received a B.S. in Architectural Engineering (cum laude) from the University of Miami in 1954, and an M.S. in Planning from the School of Architecture at Columbia University in 1956. He has earned many honors and awards, and is a member of Tau Beta Pi (National Engineering Honor Society), Pi Mu Epsilon (National Mathematics Honor Society), and other professional societies both in the United States and abroad. His biography has been published in Marquis' *Who's Who in Finance and Industry, Who's Who in the East, Who's Who in the World,* and *Dictionary of International Biography,* and he has been the winner of several architectural and master planning competitions.

   **Professor Dadras** has worked as a designer with several outstanding architects from 1950 to 1964 on such distinguished projects as the House of Seagram, the Daily News Building, the Time and Life Building, the Metropolitan Opera at Lincoln Center, and the Chrysler Pavilion at the 1963–64 New York World's Fair. He has been practicing in his own firm, Dadras International (Architects, Engineers, Planners), for the past 30 years. He and his wife, Ursula, live in Douglas Manor, New York, where they are visited often by their four children and two grandchildren

## OTHER RELATED McGRAW-HILL BOOKS
## BY ALY S. DADRAS

*Mechanical Systems for Architects*

Forthcoming books:

*HVAC Systems for Architects*

# Preface

This book is written with an understanding of the feelings of architects, architectural designers, and architectural students, who are so much involved in design and drawings and whose artistic enthusiasm does not allow them to get involved in technical and research problems which are complicated and difficult to follow.

This book is written in such a way that it is easy to follow and understand. All answers to questions are readily available. All problems are solved step by step, using the American system of measurements in order to eliminate confusion. All charts, figures, pictures, and tables are simply referred to as "Figure or Fig." in order to be identified readily.

General information given in Part One will aid a better understanding of the subjects throughout this book.

## WHY ARCHITECTS AND ARCHITECTURAL STUDENTS MUST BE KNOWLEDGEABLE IN THIS FIELD

Based on my 30 years of experience in education and practice, architects, designers, and architectural students must be knowledgeable regarding how mechanical and electrical systems are designed, calculated, and placed in structures, because approximately 42 percent of the total budget of a project is allocated for mechanical and electrical equipment, and the architect is directly responsible to the owner for the work performed by his or her consultant(s). Therefore, architects *must* be able to check the design and calculations done by their consultant(s) in order to protect themselves as well as their clients.

**It is a matter of survival in the professional practice.**

In a small project, the architect should be able to design and calculate and produce the construction documents for the mechanical and electrical equipment, thus earning the fee otherwise given to a consultant(s).

**Profit is made by knowing the field.**

Furthermore, architects and architectural students cannot begin to design a successful building(s) without knowing the space requirements for toilet facilities, kitchen equipment, vertical transportation, mechanical rooms, and the effect of mechanical, electrical, and H.V.A.C systems on their structure(s).

**Successful design cannot be accomplished without knowledge of the field.**

All information and calculations, etc., given in this book are based on the American system of measurements. Tables of conversion are provided for converting the American system to the metric system. It is suggested that all calculations be performed using the American system of measurements in the United States, and the following procedures be followed if the work is to be done for the regions or the countries where the metric system is in practice:

1. Convert all dimensions and other requirements from the metric system to the American system by using Fig. G-11.
2. Use these figures to solve the problems.
3. When you solve the problem using the American system, convert the solution to the metric system by using Fig. G-12.

*Aly S. Dadras*

# Acknowledgments

I would like to give my sincere thanks and appreciation to:

**Dr. Matthew Schure, Ph.D.**

President

New York Institute of Technology

**Dr. King V. Cheek, J.D., L.C.D., L.H.D.**

Vice President of Academic Affairs

New York Institute of Technology

**Mr. Joel E. Stein**

Senior Editor, Architecture

McGraw-Hill, Inc.

Professional Book Group

**Mrs. Ursula M.S. Dadras, B.F.A., M.P.S.**

Editor of this book

for the devoted guidance, moral support, and encouragement which they have given me in writing and preparing this book.

**Prof. Victor S. Dadras, B.Arch., M.Arch.U.D., R.A.**

**Prof. Robert S. Dadras, B.Arch., R.A.**

for their technical assistance.

*Aly S. Dadras*

# CONTENTS

# Part 1

## GENERAL INFORMATION

# G

## General Information

# CONTENTS

## Part 2

# E

## ELECTRICITY

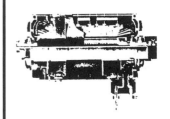

# Generating Electricity

# Electric Systems Components and Calculations

# CONTENTS

# Part 3

## ELECTRICAL SYSTEMS

# S

## Electrical Systems

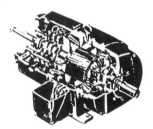

## Electrical System Voltage

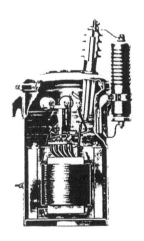

## Transformer System

# Electrical Motors

# CONTENTS

## Part 4

# ELECTRICAL CONDUCTORS (WIRING)

# C

## Electrical Conductors (Wiring)

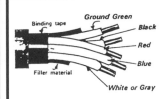

*CABLE*

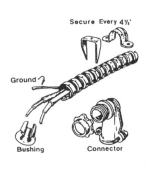

# CONTENTS

## Part 5

## ELECTRIC SERVICE TO BUILDING(S)

# B

## Electric Service to Building(s)

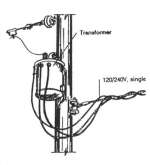

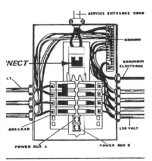

*A FLUSH-MOUNTED PANELBOARD*

# CONTENTS

**Part 6**

## ELECTRICAL WIRING DESIGN

### Electrical Wiring Design

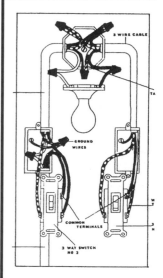

### Branch Circuit Design

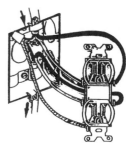

# CONTENTS

# Part 7

**L**

## LIGHTING SYSTEMS FOR BUILDING(S)

### Introduction to Lighting

### Lighting Systems for Building(s)

# Electric Light Sources

# Lighting Design and Calculation

## Local Lighting Calculation

## General Lighting Calculation

# CONTENTS

# Part 8

# VERTICAL TRANSPORTATION SYSTEMS

# V

## Vertical Transportation Systems

## Elevators

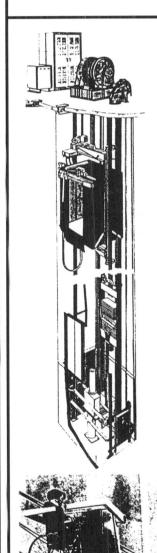

# Elevator Selection

# Vertical Transportation Systems for Handicapped Persons

# Escalators and Moving Ramps

# References

I would like to give my appreciation and acknowledgment to the following who have kindly provided photographs, drawings, and tables which supplement this book.

Acme Electric Corporation

American National Standards Institute, Inc.

Arco Solar Incorporated

The Cheney Company

Dadras International (Architects, Engineers, Planners)

Ewing Galloway

G.E. Lighting

General Electric Company

Gillespie Corporation

Illuminating Engineering Society of North America (IES)

*The Illustrated Columbia Encyclopedia*

Ked Company (Kedco)

LaMarche Manufacturing Company

McGraw-Hill Information Systems Co., Sweet's Division

National Electrical Code (NEC)

Nema Standards

Onan Corporation

Osram Sylvania Incorporated "Sylvania"

Otis Elevator Company

Siemens Solar Corp. (Arco Solar)

SKG Lifts Incorporated

Square D. Company

*Travel Magazine*

U.S. Dept. of Housing and Urban Development

Walker Systems, Incorporated

Westinghouse Electric Corporation

Westmont Industries

Wiremold Company

# CONTENTS

Part 1

## GENERAL INFORMATION

G

## General Information

# GENERAL INFORMATION

## G-1 POWER GENERATION (Fig. G-1)

The prime movers **turbine** and **generator** are the basic components of the

### generation of electricity

### TURBINE

A turbine is a rotating engine that converts the forces of fuels, water, and wind, etc., into

### mechanical energy

capable of rotating a shaft which is connected to a generator.

### GENERATOR

A generator is a machine used to change mechanical energy into

### electrical energy

This is done by induction, which **Michael Faraday** discovered in 1821.

"When a conductor is passed through a magnetic field, a current is forced through it, or induced. "

The generator is simply a mechanical arrangement for rotating an open coil of wire between the poles of a permanent magnet, leading the current at a certain voltage to an external circuit.

### ELECTRICITY

The modern electron theory states that "the fundamental nature of all matter is electrical." The atom, the smallest unit of an element, is made up of negative particles called **electrons,** positive particles named **protons,** and electrically neutral particles called **neutrons.** In a normal atom, the charges of the protons and electrons balance; if an electron leaves an atom, the atom becomes positively charged, and if an electron is added, the atom becomes negatively charged.

A body can be charged with electricity when it is touched by a charged body, and it is called **conductor.** The movement of electrons in a conductor is called **current of electricity.** The materials that do not pass on an electric charge readily are called **insulation.**

## G-2  ELECTRIC ENERGY

There are two types of electric energy produced in the United States today:

1. **Direct current (DC or dc)**
2. **Alternating current (AC or ac)**

1. **Direct current** is produced by a DC generator and is utilized for special equipment requiring exact revolutions per minute (RPM) for proper operation (vertical transportation and special compressors and motors, etc.).

   Small quantities of direct current are produced by batteries* or by rectifiers for emergency lighting, communication, signals, and control equipment, etc.

   * All batteries are charged by direct current.

2. **Alternating current** is produced by an AC generator, which is commonly referred to as an *alternator.*

   The bulk of electrical energy used today is in the form of alternating current.

**4**

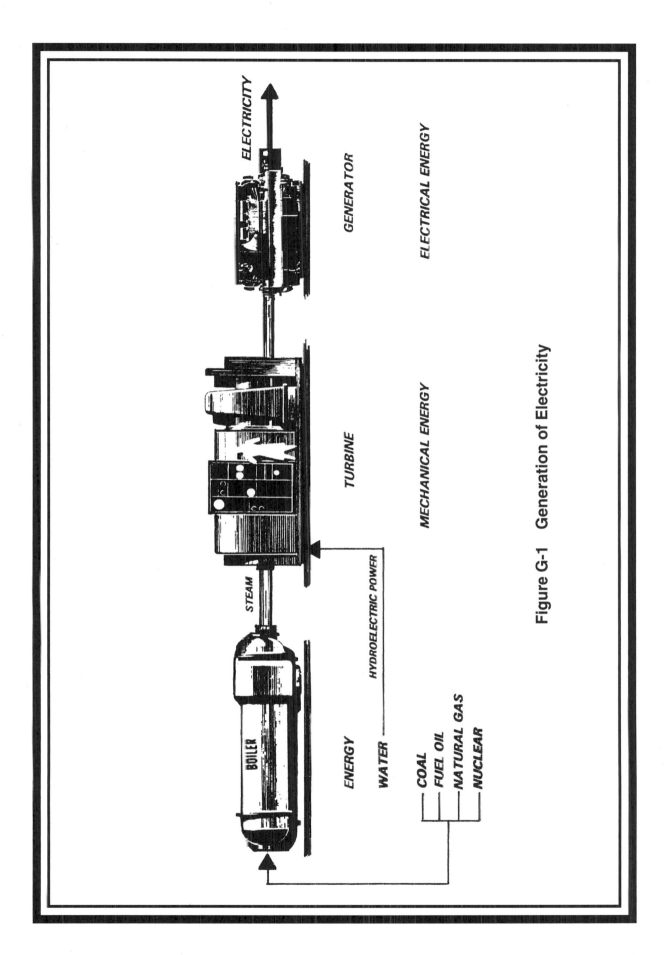

Figure G-1   Generation of Electricity

| | |
|---|---|
| **600 B.C.** | It was discovered that by rubbing amber it became charged. |
| **1600 A.D.** | **William Gilbert** described the electrification of many substances. Gilbert is called the father of modern electricity. |
| **1660 A.D.** | **Otto von Guericke** invented a crude machine for producing static electricity. |
| **1729 A.D.** | **Stephen Gray** distinguished between conductors and nonconductors. |
| | **D. F. du Fay** recognized two kinds of electricity. |
| **1752 A.D.** | **Benjamin Franklin** and **Ebenezer Kinnersley** named these two kinds of electricity "positive" and "negative." |
| **1745 A.D.** | **Pieter van Musschenbroek** invented the Leyden jar which stored static electricity. |
| **1747 A.D.** | **William Watson** discharged a Leyden jar through a circuit, and comprehension of the current and circuit started a new field of experimentation. The invention of the battery opened an interest in currents. |
| **1827 A.D.** | Continuous current from batteries was the basis for the discovery of **G. S. Ohm**'s laws and |
| **1841 A.D.** | **J. P. Joule**'s law of electrical heating. **Ohm**'s laws together with **G. R. Kirchoff**'s rules, which came later, are the basic means of making circuit calculations. |
| **1819 A.D.** | **Hans Christian Oersted** discovered that a magnetic field surrounds a wire-carrying current. |
| **1821 A.D.** | **André-Marie Ampère** had established several electromagnetic laws. |
| | **D. F. Arago** invented the electromagnet. |
| | **Michael Faraday** devised a crude form of electric motor. |
| **1831 A.D.** | **Michael Faraday** and **Joseph Henry** invented the electric generator. |
| | **Hippolyte Pixii** constructed a hand-driven model of a generator. |
| **1858 A.D.** | First steam-driven generator put into service. |

**6**

| 1878 A.D. | **C. F. Brush** installed the first arc lights in Wanamaker's store in Philadelphia. |
|---|---|
| | **Joseph W. Swan** invented the carbon filament lamp. |
| 1879 A.D. | **Thomas Alva Edison** advocated *direct current* (DC). |
| 1882 A.D. | The first central electric-light power plant in the world, "Pearl Street Plant," was completed by Thomas Edison in New York City. |
| 1888 A.D. | **C. A. Parsons** introduced a steam turbine to drive an alternator to produce *alternating current* (AC). |
| 1893 A.D. | **Nikola Tesla** designed the first hydroelectric power station in the world at Niagara Falls, producing alternating current. |
| 1900 A.D. | In the late 19th century the major fuel used for the production of electricity was coal. |
| 1950 A.D. | By 1950 clean-burning fuels of natural gas and oil, a product of petroleum, had taken the lead for producing electricity. |
| | Since 1950, nuclear power has been playing a role in producing electricity. |

*Thomas Alva Edison, photographed in 1915 at his laboratory in West Orange, N.J.*

*Thomas A. Edison's Menlo Park, N.J. laboratory when the first successful incandescent electric light bulb was turned on, Oct. 19, 1879.*

*Edison's first incandescent lamp. The filament of carbonized cotton sewing thread burned for 40 hours.*

THE ILLUSTRATED COLUMBIA ENCYCLOPEDIA

8

**G-4    THE FATHER OF MODERN ELECTRICITY**

**William Gilbert** or Gilberd (1540–1603) was an English physician and scientist especially noted for his studies of magnetism. He discovered the unit for magnetomotive force = 0.7958 ampere-turn.

Also, he described the electrification of many substances. As a result, Gilbert is called "the father of modern electricity."

**G-5    THOMAS ALVA EDISON (1847–1931)**

He advocated *direct current,* and created the first commercially practical incandescent lamp in 1879 (Fig. G-2). He developed a complete electrical distribution system for light and power, including generators, motors, light sockets, junction boxes, safety fuses, underground conductors, and other devices (Figs. G-3 and G-4). Although he held over 1300 U.S. and foreign patents, his greatest achievement was the **Pearl Street Plant** (1882) in New York City, **the first central electric-light power plant in the world** producing direct current.

**G-6    PARSON AND FERRANTI**

**C. A. Parson** in 1888 introduced a steam turbine to drive an alternator, producing *alternating current.*

**S. Z. Ferranti** designed a power station in England which proved that high-voltage alternating current could be transmitted more efficiently. Alternating current won out in the United States in 1893 when **Nikola Telsa** designed the first hydroelectric power plant in the world at Niagara Falls, producing *alternating current.*

**G-7    NIKOLA TESLA (1856–1943)**

Tesla was a Yugoslavian physicist who came to the United States in 1884. He was a pioneer in the field of high-tension electricity (alternating current). He made many discoveries and inventions of great value to the development of radio transmission and the field of electricity. These include (Fig. G-5):

*Edison's first patented dynamo (a generator to produce electric current) was invented in 1875 producing ¾ kW.*

**Figure G-3**

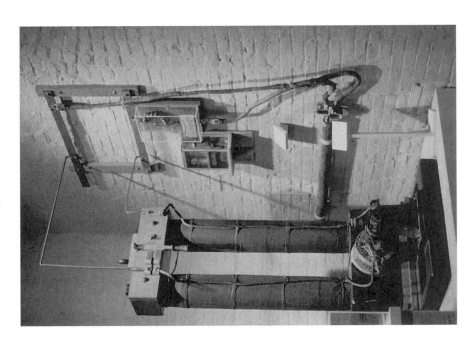

*A 90% efficient dynamo producing 110V was devised so that a multiple circuit of lights could be constantly lit.*

10

A distribution cable was also devised so to serve the sources tested.

**Figure G-4**

Edison also invented a device to measure the output of his machine. This "dynamometer" is the equivalent of today's electric meter.

The ill-fated Wardenclyffe tower built in 1901–03. It was intended for radio broadcasting and wireless transmission of power across the Atlantic.

**Figure G-5**

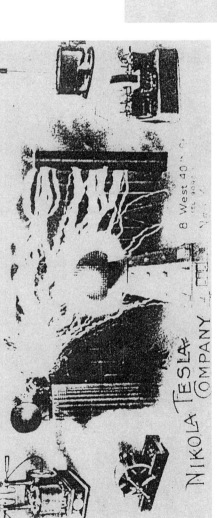

The letterhead of Tesla's business stationery recalls some of his more important inventions.

*NIKOLA TESLA*

A system of arc lighting

The Tesla induction motor

System of alternating-current transmission

The Tesla coil

Generators of high-frequency current

Transformer to increase oscillating currents to high potentials

System of wireless communication (radio) (Eight months after his death, the U.S. Supreme Court confirmed Tesla—*not* Marchese Marconi—as the inventor of the "wireless radio.")

System of transmitting electric power without wires

In partnership with **George Westinghouse,** Tesla designed the first hydroelectric power station in the world at Niagara Falls in 1893. From this time on, alternating current won out in the United States and the world. After giving so much to so many throughout the world, he lived his final years in near poverty. For his accomplishments in the field of electricity he should be recognized as **"the father of alternating current."**

## G-8    BENJAMIN FRANKLIN (1706–1790)

American statesman, scientist, and writer (Fig. G-6). His discoveries won him recognition from the leading scientists. They include:

1.  His spectacular experiment of flying a kite in a thunderstorm proved the identity of electricity in lightning.
2.  His invention of the lightning rod was among the series of investigations that won him recognition from the leading scientists.
3.  After C. F. du Fay recognized the existence of two kinds of electricity, Benjamin Franklin and Ebenezer Kinnersley named the two kinds of electricity *positive* and *negative.*
4.  He had many discoveries and inventions in the field of electricity.

*JAMES WATT*

**Figure G-8**

*ANDRÉ M. AMPÈRE*

**Figure G-7**

*BENJAMIN FRANKLIN*

**Figure G-6**

THE ILLUSTRATED COLUMBIA ENCYCLOPEDIA

14

**G-9    GEORG S. OHM (1787–1854)**

**OHM'S LAW**

Ohm was a German physicist. His study of electric current led to his formulation of the law known as "**Ohm's law.**"

Ohm's law states that when an electromotive force (volt $V$) is applied to a conductor, the current throughout it (amperes $I$) is equal to the volt $V$ divided by resistance $R$ (in ohm) of the conductor.

$$I = \frac{V}{R}$$

Ohm's law implies that the ratio of electromotion force to the current is a constant, and is independent of the current strength.

**G-10    ANDRÉ-MARIE AMPÈRE (1775–1836)**

**AMPERE (UNIT OF CURRENT)**

Ampère was a French physicist and mathematician (Fig. G-7).

He is known for his contributions to electrodynamics, formulation of **Ampère's law,** and invention of a static needle.

**Ampere, the electrical unit of current strength** and **unit of measurement of rate of flow of electricity,** is named after him.

**G-11    ALESSANDRO G. VOLTA (1745–1827)**

**VOLT (UNIT OF POWER)**

Volta was an Italian physicist. He invented voltaic pile (Volta's pile), electric condenser, and voltaic cell.

**Unit of measurement of the electromotive force,** commonly known as emf, is named after him.

**G-12   JAMES WATT (1736–1819)**

**WATT (UNIT OF POWER)**

He patented his steam engine in 1869, and his discoveries include:

1.   A separate condenser, which caused 75 percent saving in fuel

2.   Use of steam pressure to move the piston in both directions (reciprocal action)

3.   A crank flywheel mechanism

4.   A mercury steam gauge and a governor

All of his discoveries prepared the steam engine for the start of **the Industrial Revolution.**

**Watt, the unit of power,** is named after him (Fig. G-8).

There are two units of power: (1) horsepower and (2) watt.

1.   *Horsepower* (HP) is based on the concept that a horse can do 550 foot-pounds of

work per second.

Foot-pound is when 1 pound of weight is moved 1 foot.

**746 watt is equal to 1 horsepower (HP)**

2.   *Watt* is the unit of electrical power. It is defined as the power necessary to maintain a

current of 1 ampere under a pressure of 1 volt.

**Watt (W) = 1 V**

Electrical power is sold by kilowatt hour.

**Kilowatt = 1000 W**

One watt is equal to 1 joule per second.

**G-13    JAMES P. JOULE (1818–1889)**

**JOULE (UNIT OF WORK)**

Joule was an English physicist who invented an electromagnetic engine, which was a great contribution to the field of electricity, heat, and thermodynamics.

**The electrical unit of work** is named after him.

**Joule is the amount of work required in 1 second to maintain a current of 1 ampere in a resistance of 1 ohm.**

Joule = 10,000,000 ergs (the unit of work in the cgs* system)

Joule = 0.7376 foot-pounds

Joule = 0.1020 kilogram-meters

Joule = 0.24 calories (15°)

Joule = 0.0009487 Btu

* cgs stands for centimeter-gram-second.

**G-14    HEINRICH RUDOLF HERTZ (1857–1894)**

Hertz was a German physicist who discovered electromagnetic waves known as *hertzian waves* and *radio waves.*

He demonstrated that these waves are long, transverse waves that travel at the velocity of light (186,000 miles per second) and can be

1.    Reflected
2.    Refracted and
3.    Polarized like light

He also investigated the electric discharge in rarefied gases.

**Cycles per second in alternating current** is named after him.

**Cycle/sec = hertz (Hz)**

# AMERICAN AND BRITISH SYSTEM

## Linear Measure

12 inches = 1 foot
3 feet = 1 yard
5½ yards = 1 rod
220 yards = 1 furlong
5,280 feet = 1 mile
6 feet = 1 fathom
6,076.1 feet = 1 nautical mile

## Square Measure

144 square inches = 1 square foot
9 square feet = 1 square yard
30¼ square yards = 1 square rod
160 square rods = 1 acre
4,840 square yards = 1 acre
640 acres = 1 square mile

## Cubic Measure

1,728 cubic inches = 1 cubic foot
27 cubic feet = 1 cubic yard

## Liquid Measure

16 ounces = 1 (U.S.) pint
20 ounces = 1 imperial (British) pint
2 cups = 1 pint
2 pints = 1 quart
4 quarts = 1 gallon
1.2 U.S. gallons = 1 imperial (British) gallon

## Weights: Avoirdupois

16 drams = 1 ounce
16 ounces = 1 pound
112 pounds = 1 long hundredweight
2000 pounds = 1 short ton
2240 pounds = 1 long ton

## Weights: Troy and Apothecary

480 grains = 1 ounce
12 ounces = 1 pound

**Figure G-9**

## G-15   WEIGHTS AND MEASURES:

The crude versions of the weights and measures probably date back to prehistoric times. Early units were commonly based on body measurements and on plant seeds.

As civilization progressed, technological and commercial requirements led to increased standardization.

Today the chief systems are:

   a.   *American and British System* (Fig. G-9)
   b.   *Metric System* (Fig. G-10)

**a.   American and British System**

This system evolved in England and was carried to America by English colonists, and in 1856 became official weights and measures in the United States.

In the American and British System, two sets of weights are employed:

   1.   *Avoirdupois.*   Used in general commerce. It was legalized in England in 1303, and is based on 16 ounces per pound.
   2.   *Troy.*   Used for weighing precious metals. It was legalized in England in 1527, and is based on 12 ounces per pound.

**b.   Metric System**

This is a system of measurements and weights planned in France and adopted there in accordance with a law passed in 1779. It was subsequently adopted by many nations. The use of the metric system was permitted in England, and the U.S. government permitted the use of the metric system in 1866.

The U.S. Metric Conversion Act (Public Law 94-168) of 1975 prescribed a gradual voluntary conversion to the metric system (Système International d'Unités (S1).

# METRIC SYSTEM

### Linear Measure

10 millimeters = 1 centimeter
10 centimeters = 1 decimeter
10 decimeters = 1 meter
10 meters = 1 dekameter
1,000 meters = 1 kilometer

### Square Measure

100 square millimeters = 1 square centimeter
100 square centimeters = 1 square decimeter
10,000 square centimeters = 1 square meter
100 square meters = 1 are
100 ares = 1 hectare
10,000 ares = 1 square kilometer

### Cubic Measure

1,000 cubic millimeters = 1 cubic centimeter
1,000 cubic centimeters = 1 cubic decimeter
1,000 cubic decimeters = 1 cubic meter

### Liquid Measure

10 milliliters = 1 centiliter
100 centiliters = 1 liter
1,000 liters = 1 kiloliter

### Weights

10 milligrams = 1 centigram
10 centigrams = 1 decigram
10 decigrams = 1 gram
100 centigrams = 1 gram
10 grams = 1 dekagram
10 dekagrams = 1 hectogram
10 hectogram = 1 kilogram
1,000 grams = 1 kilogram
1,000 kilograms = 1 ton

**Figure G-10**

## G-16    CONVERSION FACTORS

All information and calculations, etc., given in this book are based on the American system of measurement. Tables of conversion are provided for converting the American system to the metric system. It is suggested that all calculations be performed using the American system, which is commonly used in the United States. If the work is to be done in the countries or regions where the metric system is in practice, proceed as follows:

1.    **Convert all dimensions and other requirements from the metric system to the American system using Fig. G-11.**

2.    **Use these figures to solve the problems.**

3.    **When you solve the problem using the American system, convert the solution to the metric system by using Fig. G-12.**

# CONVERSION FACTORS
## METRIC SYSTEM TO AMERICAN SYSTEM

| | | | | |
|---|---|---|---|---|
| **Length** | millimeter | mm $\times$ 0.0393 | = inch | in |
| | meter | m $\times$ 3.2808 | = foot | ft |
| | meter | m $\times$ 1.0936 | = yard | yd |
| | kilometer | km $\times$ 0.6213 | = mile U.S. | mi |
| **Area** | sq. millimeter | mm$^2$ $\times$ 0.00155 | = sq. inch | in$^2$ |
| | sq. meter | m$^2$ $\times$ 10.7639 | = sq. foot | ft$^2$ |
| | sq. meter | m$^2$ $\times$ 1.1959 | = sq. yard | yd$^2$ |
| | sq. kilometer | km$^2$ $\times$ 0.3861 | = sq. mile | mi$^2$ |
| | sq. meter | m$^2$ $\times$ 0.000247 | = acre | A |
| | hectare | h $\times$ 2.471 | = acre | A |
| **Volume** | cubic millimeter | mm$^3$ $\times$ 0.000061 | = cubic inch | cu. in. |
| | cubic meter | m$^3$ $\times$ 35.3146 | = cubic foot | cu. ft. |
| | cubic meter | M$^3$ $\times$ 1.3079 | = cubic yard | cu. yd. |
| | liter | l $\times$ 0.2641 | = gallon U.S. | gal |
| | liter | l $\times$ 1.0566 | = quart U.S. | qt |
| **Mass** | gram | g $\times$ 0.03527 | = ounce (avo) | oz |
| | kilogram | kg $\times$ 2.2046 | = pound (avo) | lb |
| | kilogram | kg $\times$ 0.001102 | = short ton | t |
| **Pressure** | kilopascal | KPa $\times$ 0.145 | = pound-force/m$^2$ psi | |
| | kilopascal | KPa $\times$ 0.3345 | = foot of water 1 psi | |
| | kilopascal | KPa $\times$ 0.2953 | = in. of mercury-32°F | |
| **Power** | watt | W $\times$ 0.7375 | = ft.lb.f/s | |
| | watt | W $\times$ 3.4121 | = Btuh | |
| | kilowatt | kW $\times$ 1.341 | = HP (550 ft-lbf/s) | |
| **Angle** | radian | rad $\times$ 57.2957 | = degree | deg |
| **Temperature Celsius** | | C $(1.8\times°C + 32)$ | = degree Fahrenheit | |

**Figure G-11**

# CONVERSION FACTORS
## AMERICAN SYSTEM TO METRIC SYSTEM

| | | | | | |
|---|---|---|---|---|---|
| **Length** | inch | in | × 25.4 | = millimeter | mm |
| | foot | ft | × 0.3048 | = meter | m |
| | yard | yd | × 0.9144 | = meter | m |
| | mile U.S. | mi | × 1.6093 | = kilometer | km |
| **Area** | sq. inch | in$^2$ | × 645.16 | = sq. millimeter | mm$^2$ |
| | sq. foot | ft$^2$ | × 0.0929 | = sq. meter | m$^2$ |
| | sq. yard | yd$^2$ | × 0.8361 | = sq. meter | m$^2$ |
| | sq. mile U.S. | mi$^2$ | × 2.5899 | = sq. kilometer | km$^2$ |
| | acre | A | × 4046.873 | = sq. meter | m$^2$ |
| | acre | A | × 0.4046 | = hectare | h |
| **Volume** | cubic inch | in$^3$ | × 16387.06 | = cubic millimeter | cu. mm |
| | cubic foot | ft$^3$ | × 0.0283 | = cubic meter | cu. m |
| | cubic yard | yd$^3$ | × 0.7645 | = cubic meter | cu. m |
| | gallon U.S. | gal | × 3.7854 | = liter | l |
| | quart U.S. | qt | × 0.9463 | = liter | l |
| **Mass** | ounce (avo) | oz | × 28.3495 | = gram | g |
| | pound (avo) | lb | × 0.4535 | = kilogram | kg |
| | short ton | t | × 907.185 | = kilogram | kg |
| **Pressure** | pound-force/m$^2$ | psi | × 6.8947 | = kilopascal | KPa |
| | foot of water | 1 psi | × 2.9889 | = kilopascal | KPa |
| | inch of mercury | (32°F) | × 3.3863 | = kilopascal | KPa |
| **Power** | ft-lb-f/s | | × 1.3558 | = watt | W |
| | Btuh | | × 0.293 | = watt | W |
| | HP (550 ft. lb/s) | | × 0.7457 | = kilowatt | kW |
| **Angle** | degree | deg | × 0.01745 | = radian | +ad |
| **Temperature Fahrenheit** | | F | (°F − 32 ÷ 1.8) | = degree Celsius | C |

**Figure G-12**

# G-17   ELECTRICAL SYMBOLS

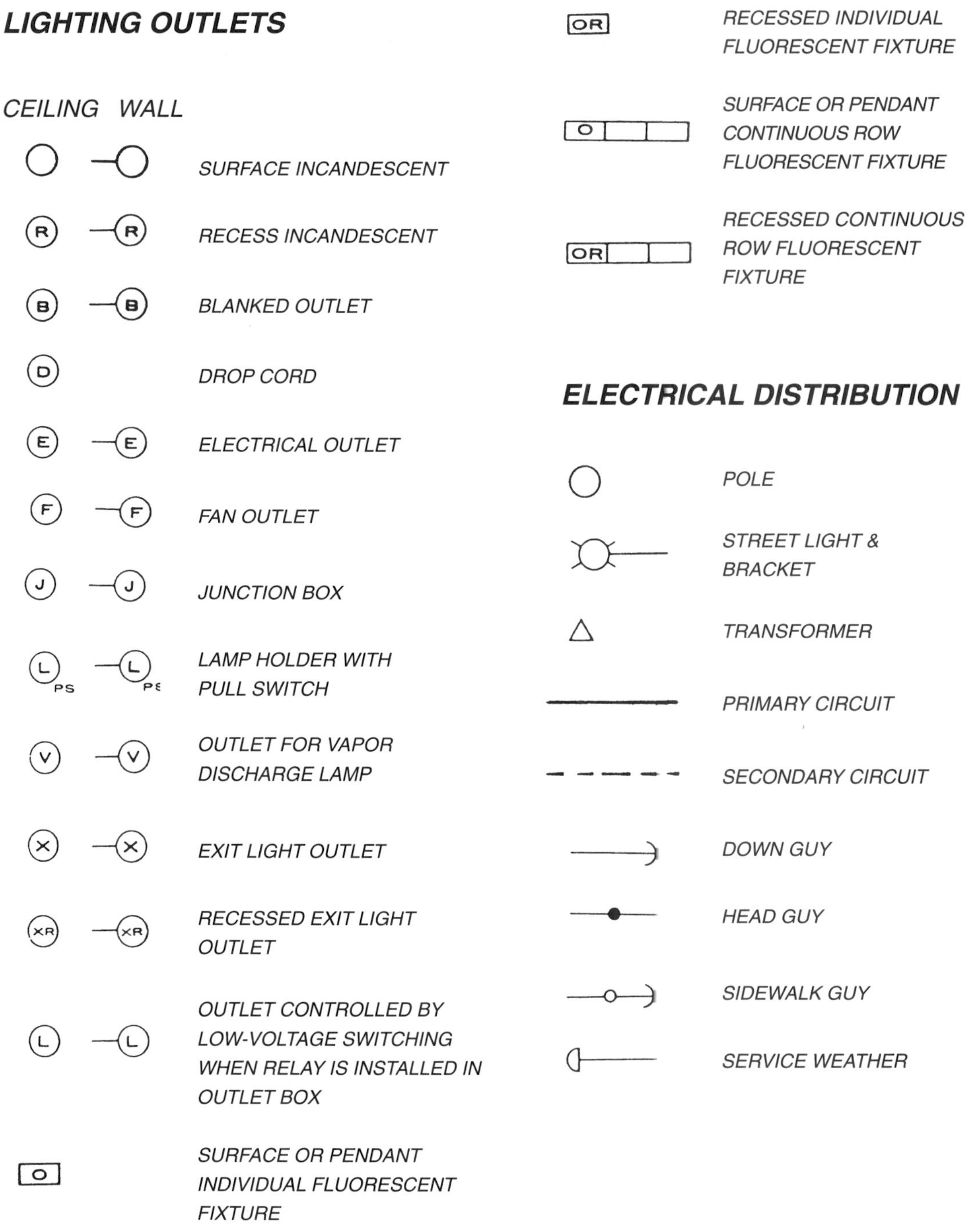

## LIGHTING OUTLETS

CEILING   WALL

| | | |
|---|---|---|
| ○ | ─○ | SURFACE INCANDESCENT |
| Ⓡ | ─Ⓡ | RECESS INCANDESCENT |
| Ⓑ | ─Ⓑ | BLANKED OUTLET |
| Ⓓ | | DROP CORD |
| Ⓔ | ─Ⓔ | ELECTRICAL OUTLET |
| Ⓕ | ─Ⓕ | FAN OUTLET |
| Ⓙ | ─Ⓙ | JUNCTION BOX |
| Ⓛ PS | ─Ⓛ PS | LAMP HOLDER WITH PULL SWITCH |
| Ⓥ | ─Ⓥ | OUTLET FOR VAPOR DISCHARGE LAMP |
| ⓧ | ─ⓧ | EXIT LIGHT OUTLET |
| (xR) | ─(xR) | RECESSED EXIT LIGHT OUTLET |
| Ⓛ | ─Ⓛ | OUTLET CONTROLLED BY LOW-VOLTAGE SWITCHING WHEN RELAY IS INSTALLED IN OUTLET BOX |
| ▭O | | SURFACE OR PENDANT INDIVIDUAL FLUORESCENT FIXTURE |

| | |
|---|---|
| OR | RECESSED INDIVIDUAL FLUORESCENT FIXTURE |
| O▢▢ | SURFACE OR PENDANT CONTINUOUS ROW FLUORESCENT FIXTURE |
| OR▢ | RECESSED CONTINUOUS ROW FLUORESCENT FIXTURE |

## ELECTRICAL DISTRIBUTION

| | |
|---|---|
| ○ | POLE |
| ⊗ | STREET LIGHT & BRACKET |
| △ | TRANSFORMER |
| ────── | PRIMARY CIRCUIT |
| ─ ─ ─ ─ | SECONDARY CIRCUIT |
| ───→ | DOWN GUY |
| ──●─ | HEAD GUY |
| ─○─→ | SIDEWALK GUY |
| ⊂──── | SERVICE WEATHER |

**Figure G-13   Electrical Symbols**

24

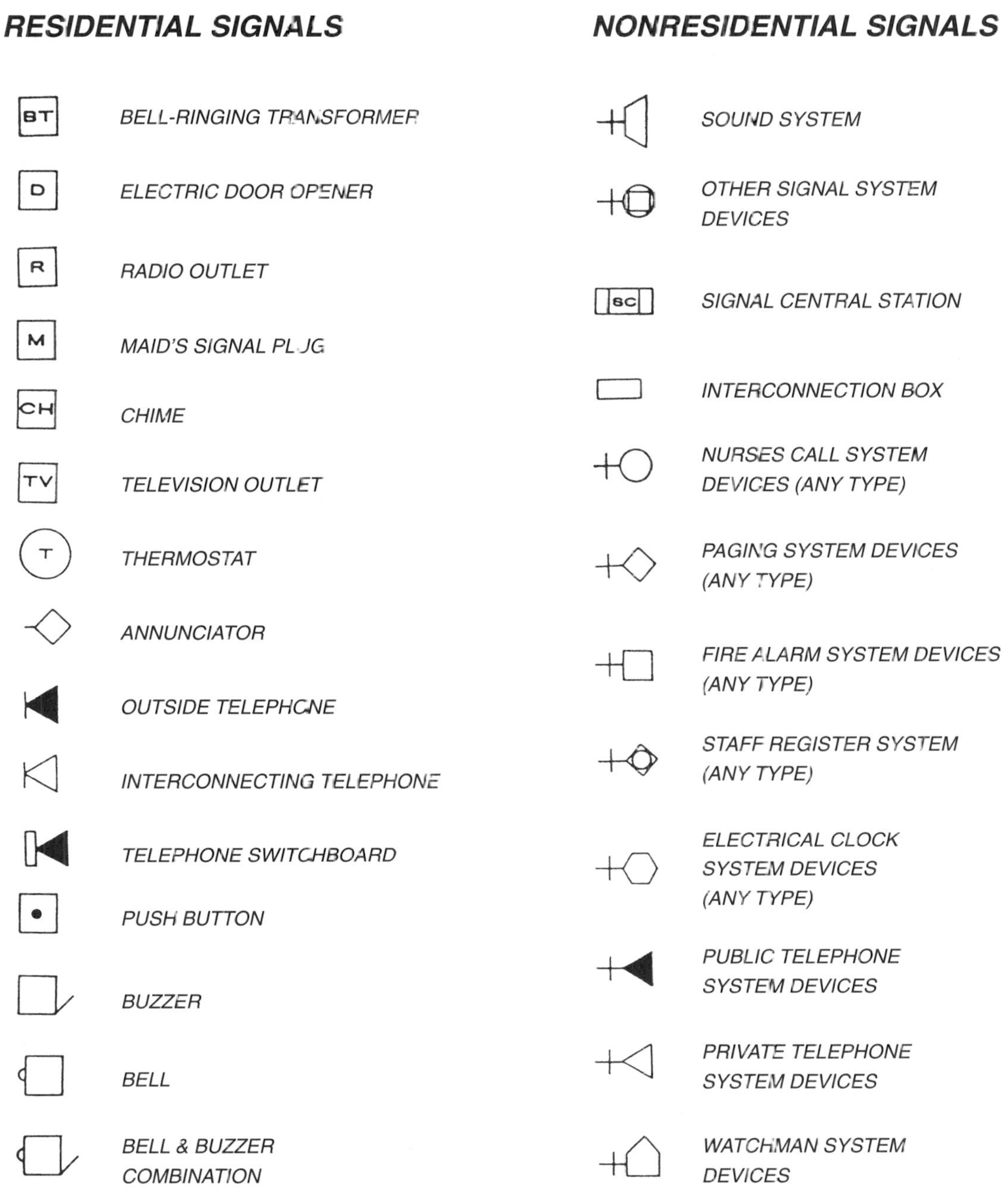

## RESIDENTIAL SIGNALS

| | |
|---|---|
| BT | BELL-RINGING TRANSFORMER |
| D | ELECTRIC DOOR OPENER |
| R | RADIO OUTLET |
| M | MAID'S SIGNAL PLUG |
| CH | CHIME |
| TV | TELEVISION OUTLET |
| T | THERMOSTAT |
| | ANNUNCIATOR |
| | OUTSIDE TELEPHONE |
| | INTERCONNECTING TELEPHONE |
| | TELEPHONE SWITCHBOARD |
| | PUSH BUTTON |
| | BUZZER |
| | BELL |
| | BELL & BUZZER COMBINATION |

## NONRESIDENTIAL SIGNALS

| | |
|---|---|
| | SOUND SYSTEM |
| | OTHER SIGNAL SYSTEM DEVICES |
| SC | SIGNAL CENTRAL STATION |
| | INTERCONNECTION BOX |
| | NURSES CALL SYSTEM DEVICES (ANY TYPE) |
| | PAGING SYSTEM DEVICES (ANY TYPE) |
| | FIRE ALARM SYSTEM DEVICES (ANY TYPE) |
| | STAFF REGISTER SYSTEM (ANY TYPE) |
| | ELECTRICAL CLOCK SYSTEM DEVICES (ANY TYPE) |
| | PUBLIC TELEPHONE SYSTEM DEVICES |
| | PRIVATE TELEPHONE SYSTEM DEVICES |
| | WATCHMAN SYSTEM DEVICES |

**Figure G-14   Electrical Symbols**

## RECEPTACLE OUTLETS

|  |  |
|--------|-------------|
| ─⊖ | SINGLE-RECEPTACLE OUTLET |
| ═⊖ | DUPLEX-RECEPTACLE OUTLET |
| ═⊕ | TRIPLEX-RECEPTACLE OUTLET |
| ─⊕ | QUADRUPLEX-RECEPTACLE OUTLET |
| ─◒ | DUPLEX-RECEPTACLE OUTLET— SPLIT WIRED |
| ─◒ | TRIPLEX-RECEPTACLE OUTLET— SPLIT WIRED |
| ─△ | SINGLE SPECIAL-PURPOSE RECEPTACLE OUTLET |
| ─△ | DUPLEX SPECIAL-PURPOSE RECEPTACLE OUTLET |
| ═⊖ R | RANGE OUTLET |
| ─▲ DW | SPECIAL-PURPOSE CONNECTION |
| ⊖ X" | MULTIOUTLET ASSEMBLY |
| ⓒ | CLOCK HANGER RECEPTACLE |
| Ⓕ | FAN HANGER RECEPTACLE |
| ⊟ | FLOOR SINGLE-RECEPTACLE OUTLET |
| ⊟ | FLOOR DUPLEX-RECEPTACLE OUTLET |

## SWITCH OUTLETS

|  |  |
|--------|-------------|
| S | SINGLE POLE SWITCH |
| $S_2$ | DOUBLE POLE SWITCH |
| $S_3$ | THREE-WAY SWITCH |
| $S_4$ | FOUR-WAY SWITCH |
| $S_D$ | AUTOMATIC DOOR SWITCH |
| $S_K$ | KEY-OPERATED SWITCH |
| $S_P$ | SWITCH AND PILOT LAMP |
| $S_T$ | TIME SWITCH |
| Ⓢ | CEILING PULL SWITCH |
| ─⊖$_S$ | SWITCH AND SINGLE RECEPTACLE |
| ═⊖$_S$ | SWITCH AND DOUBLE RECEPTACLE |

**Figure G-15   Electrical Symbols**

26

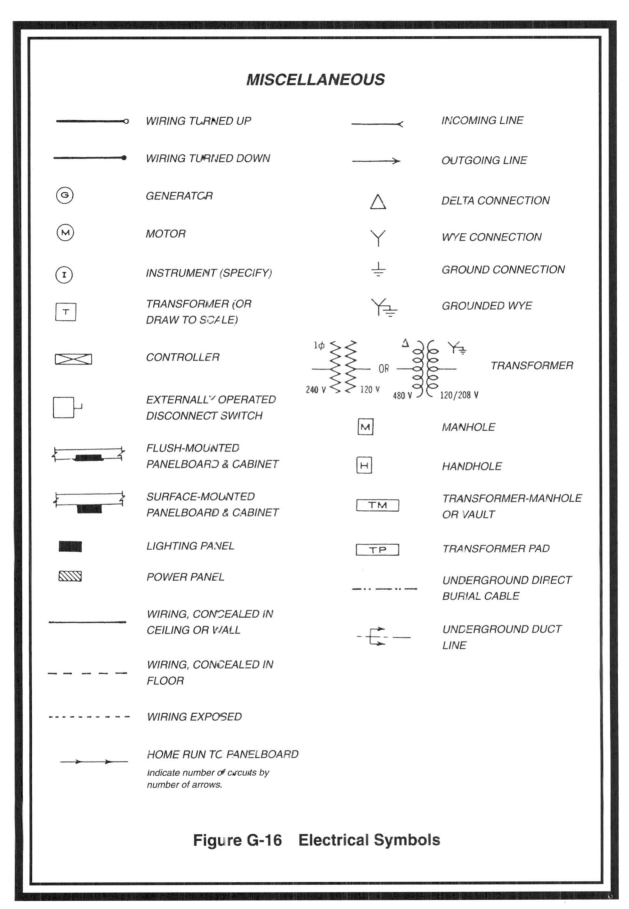

## MISCELLANEOUS

| | |
|---|---|
| ━━━━━○ WIRING TURNED UP | ━━━━━< INCOMING LINE |
| ━━━━━● WIRING TURNED DOWN | ━━━━━→ OUTGOING LINE |
| Ⓖ GENERATOR | △ DELTA CONNECTION |
| Ⓜ MOTOR | Y WYE CONNECTION |
| Ⓘ INSTRUMENT (SPECIFY) | ⏚ GROUND CONNECTION |
| T TRANSFORMER (OR DRAW TO SCALE) | Y⏚ GROUNDED WYE |
| ⊠ CONTROLLER | 1φ 240 V 120 V OR 480 V 120/208 V TRANSFORMER |
| EXTERNALLY OPERATED DISCONNECT SWITCH | |
| FLUSH-MOUNTED PANELBOARD & CABINET | M MANHOLE |
| SURFACE-MOUNTED PANELBOARD & CABINET | H HANDHOLE |
| ▬ LIGHTING PANEL | TM TRANSFORMER-MANHOLE OR VAULT |
| ▨ POWER PANEL | TP TRANSFORMER PAD |
| ━━━━━ WIRING, CONCEALED IN CEILING OR WALL | ─ ·· ─ ·· ─ UNDERGROUND DIRECT BURIAL CABLE |
| ─ ─ ─ ─ WIRING, CONCEALED IN FLOOR | UNDERGROUND DUCT LINE |
| - - - - - - WIRING EXPOSED | |
| ──────→ HOME RUN TO PANELBOARD | |

*Indicate number of circuits by number of arrows.*

### Figure G-16    Electrical Symbols

27

# G-18 ABBREVIATIONS

| | |
|---|---|
| **A** | amperes |
| **AA** | Aluminum Association |
| **AAC** | all-aluminum conductor |
| **AASC** | aluminum alloy stranded conductors |
| **AAQ** | aquastat |
| **AC, ac** | alternating current |
| **ACSR** | aluminum conductor, steel reinforced |
| **AFF** | above finished floor |
| **AIA** | American Institute of Architects |
| **AL** | aluminum |
| **ANSI** | American National Standards Institute |
| **ASME** | American Society of Mechanical Engineers |
| **ASTM** | American Society of Testing and Materials |
| **ATL** | across-the-line starter—magnetic |
| **AWG** | American Wire Gauge |
| | |
| **BATT** | battery |
| **BG** | break-glass |
| **BT** | bell transformer |
| **BX** | nickname for *armored cable* |
| | |
| **C** | conduit |
| **CATL** | combination across-the-line magnetic starter |
| **C/B** | circuit breaker |
| **C-C** | center to center |
| **CCT** | circuit |
| **CH** | chime |
| **cgs** | centimeter-gram-second |
| **CL** | centerline |
| **CT** | control transformer |
| **Cu** | copper |

| | |
|---|---|
| **DB** | door bell |
| **DC, dc** | direct current, which is unvarying in polarity |
| **DF** | drinking fountain |
| **DH** | door holder |
| **DO** | door opener |
| **DN** | down |
| **DOAL** | diameter overall |
| **DOC** | diameter over conductor |
| **DOI** | diameter over insulation |
| **DOJ** | diameter over jacket |
| **DOS** | diameter over insulation shield |
| | |
| **EC** | electrical conductor of aluminum |
| **EC** | empty conduit |
| **EEA** | Electric Energy Association |
| **EEI** | Edison Electric Institute |
| **EHV** | extra high voltage: 230–765 kV |
| **EM** | emergency |
| **EMA** | electrical moisture absorption |
| **EMI** | electromagnetic interference |
| **EL** | elevation |
| **EP** | electropneumatic |
| **EPA** | Environmental Protection Agency |
| **EPRI** | Electric Power Research Institute |
| **ERDA** | Energy Research & Development Administration |
| **ETL** | Electrical Testing Laboratory |
| **EWC** | electric water cooler |
| | |
| **F** | fuse |
| **FA** | fire alarm |
| **F-3** | fan no. 3 |
| **FC** | fan coil unit |

| | |
|---|---|
| **F, FL** | float switch |
| **FPM** | feet per minute |
| **FS** | fused switch |
| **FV** | full voltage |
| | |
| **GFCI** | ground fault circuit interrupter |
| **GFI** | ground fault interrupter |
| **GND** | ground |
| **GPM, gpm** | gallons per minute |
| | |
| **H** | humidistat |
| **HMP, HMPE** | high-molecular-weight polyethylene |
| **HOA** | hand-off-automatic selector switch |
| **H-O-A** | hand-off-automatic switch |
| **HP** | horsepower |
| | |
| **I, IA** | intrusion alarm |
| **IACS** | International Annealed Copper Standard |
| **IBEW** | International Brotherhood of Electric Workers |
| **IC** | refers to interrupting capacity of any device required to break current (switch, circuit breaker, fuse, etc.) |
| **IC** | intercom |
| **ID** | inside diameter |
| **IEC** | International Electrochemical Commission |
| **IEEE** | Institute of Electrical and Electronics Engineers |
| **IES** | Illuminating Engineering Society |
| **IIR** | isobutylene isoprene rubber |
| **IPCEA** | Insulated Power Cable Engineers Association |
| **IR** | insulation resistance |
| **IR drop** | the voltage drop across a resistance due to the flow of current |
| **IRK** | insulation dc resistance constant; a system to classify materials according to their resistance on a 1000-foot basis at 15.5°C (60°F) |

**30**

| | |
|---|---|
| **ISO** | International Organization for Standardization, which has put together the SI units |
| $I^2t$ | relating to the heating effect of a current for a specified time, under specified conditions |
| | |
| **JB** | junction box |
| | |
| **ka** | kiloampere |
| **kc** | kilocycle, use kilohertz |
| **Kg** | kilogram |
| **kHz** | kiloherz |
| **KO** | knockout; the partially cut opening in boxes, panel cabinets and other enclosures, which can be easily knocked out |
| **kVA** | kilovolts times ampere |
| | |
| **L** | line |
| **LA** | lightning arrestor |
| **LIM** | Laboratory Inspection Manual |
| **LO** | lockout |
| **LS, HS** | low speed, high speed |
| **LV** | low voltage |
| | |
| **M** | motor |
| **Mbh** | thousands of British thermal units (Btu) per hour |
| **MCC** | motor control center |
| **MCM** | conductors of sizes from 250 MCM, which stands for thousand circular mils up to 2000 MCM and larger |
| **MD** | motorized damper |
| **MER** | mechanical equipment room |
| **MH** | mounting height |
| **MFT** | thousands of feet |
| **Mho** | reciprocal of ohm |
| **MI cable** | mineral-insulated, metal-sheathed cable |

| | |
|---|---|
| **mil** | A unit used in measuring the diameter of wire, equal to 0.001 inch |
| **mm** | millimeter = 1 meter ÷ 1000 |
| **mks** | meter-kilogram-second |
| **MOM** | momentary contact |
| **MTW** | machine tool wire |
| **Mw** | megawatt: 106 watts |
| | |
| **N** | neutral |
| **N/A** | (1) not available; (2) not applicable |
| **NBS** | National Bureau of Standards |
| **NC** | nurse call (push button) |
| **NC** | normally closed |
| **NEC** | National Electrical Code |
| **NEMA** | National Electrical Manufacturers Association |
| **NFPA** | National Fire Protection Association |
| **NIC** | not in contract |
| **NL** | night light |
| **NO** | normally open |
| | |
| **OC** | overcurrent |
| **OD** | outside diameter |
| **OF** | oxygen-free |
| **OH** | overhead |
| **OL** | overload relay |
| **OC** | on center |
| | |
| **PB** | push button |
| **PC** | pull chain |
| **PE** | pneumatic electric |
| **PES** | Power Engineering Society of IEEE |
| **PF** | power factor |
| **Pi ($\pi$)** | ratio of the circumference of a circle |
| **P-P** | peak to peak |

| | |
|---|---|
| **PL** | pilot light |
| **PS** | pressure switch |
| **PSI, psi** | pounds per square inch pressure |
| **PT** | potential transformer |
| **PVC** | polyvinyl chloride; a thermoplastic insulation and jacket compound |
| | |
| **QA** | quality assurance |
| | |
| **R** | relay |
| **REA** | reversing |
| **RF** | radio frequency: 10 kGz to GHz |
| **RFI** | radio frequency interference |
| **RHC** | reheat coil |
| **ROM** | read-only memory |
| **RPM** | revolution per minute |
| **RV** | reduced voltage |
| | |
| **S** | start button—momentary contact |
| **SD** | smoke detector |
| **S, SP** | speaker, loudspeaker |
| **SP** | single pole |
| **SPDT** | single pole, double throw |
| **S/S PB** | start–stop push button |
| **SPST** | single pole, single throw |
| **SR** | starter rack |
| **SSR** | solid-state relay |
| **ST** | stop button—momentary contact |
| **SW** | switch |
| | |
| **T** | transformer |
| **T** | thermostat |
| **TC** | telephone cabinet |
| **TC** | thermocouple; time constant; timed closing |

| | |
|---|---|
| **TDR** | time delay relay; time domain reflectometer |
| **TEL** | telephone |
| **TPR** | thermal plastic rubber |
| **TV** | television |
| **TYP** | typical |
| | |
| **URD** | underground residential distribution: a single-phase cable usually consisting of an insulated conductor having a bare concentric neutral |
| **UF** | unfused |
| **UG** | underground |
| **UH** | unit heater |
| **UHF** | ultrahigh frequency: 300 MHz to 3 GHz |
| **UL** | Underwriters Laboratories |
| **UON** | unless otherwise noted |
| **UV** | undervoltage |
| | |
| **VA** | volts times amps |
| **VD** | voltage drop |
| **VOM** | volt-ohm-multimeter |
| **VP** | vaporproof |
| **VW-1** | UL rating given single-conductor cables in regard to flame resistant properties |
| | |
| **WP** | weatherproof |
| **W, WT** | watchman's tour |
| | |
| **XP** | explosion-proof |
| **XFMR** | transformer |

# G-19 DEFINITIONS

| | |
|---|---|
| **absorptance** | The ratio of light absorbed by a material to the incident light falling on it (darker objects absorb more light than lighter-colored objects). |
| **accessible** | For wiring methods: (1) not permanently closed in by the structure; (2) capable of being removed without disturbing the building. |
| **accessible** | For equipment: not guarded by locked doors, elevation, or other means. |
| **alive** | Energized; having voltage applied. |
| **alternator** | A device to produce alternating current (generator). |
| **American Wire Gauge (AWG)** | The standard used for measuring wire in the United States. |
| **ampacity** | (1) A wire's ability to carry safely, without undue heating; (2) the current-carrying capacity of equipment. |
| **ambient temperature** | Temperature of air that surrounds an object on all sides. |
| **ammeter** | An electric meter used to measure current. |
| **ampere (A)** | Unit measuring the quantity of current (in amperes). |
| **amplification** | Procedure of expanding the strength of a sound, etc. |
| **amplifier** | The process of increasing the strength of an input. |

| **amplitude** | The maximum value of a wave. |
| **anode** | (1) Positive electrode through which current enters a nonmetallic conductor; (2) negative pole of a storage battery. |
| **antenna** | A device for transmission or reception of electromagnetic waves. |
| **appliance** | Equipment designed for a particular purpose which utilizes electricity to produce heat, light, mechanical motion, etc. |
| **appliance outlet** | (1) An outlet connected to an appliance circuit; (2) an outlet box intended for direct connection to an appliance. |
| **approved** | Acceptable to the authority enforcing a particular code. |
| **arc** | A flow of current across an insulating medium. |
| **arcing time** | The time elapsing from the severance of the circuit to the interruption of current. |
| **arc resistance** | The time required for an arc to establish a conductive path in or across a material. |
| **armature** | (1) The member in which alternating voltage is generated; (2) the member which is moved by magnetic force. |
| **armor** | Metal protector for cables. |
| **askarel** | A synthetic nonflammable insulating liquid which, when decomposed by the electric arc, evolves only nonflammable gaseous mixtures. |

| | |
|---|---|
| **automatic** | Operating by its own mechanism. |
| **automatic transfer equipment** | A device to transfer a load from one power source to another. |
| **autotransformer** | Any transformer where primary and secondary connections are made to a single cell. |
| **auxiliary** | A device or equipment which aids the main device or equipment. |
| **ballast** | A device used with fluorescent and high-intensity discharge lamps to provide the necessary circuit conditions for starting and operating the light. |
| **bar** | A long, solid metal having a cross-sectional dimension of 0.375 inch or more. |
| **bare** | A conductor which is not insulated or coated. |
| **base** | One of the terminals of a transistor. |
| **base load** | The minimum load over a period of time. |
| **battery** | (1) A device which changes chemical to electrical energy; (2) used to store electricity. |
| **belt** | The outer protective nonmetallic covering of cable. |
| **belted-type cable** | A multiple-conductor cable having a layer of insulation over the core conductor assembly. |

| | |
|---|---|
| **branch circuit** | A wiring system extending beyond the final overcurrent device protecting the circuit. |
| **branch circuit—appliance** | A circuit supplying energy to one or more outlets to which appliances are to be connected. |
| **branch circuit—general purpose** | A circuit that supplies a number of outlets for lighting and appliances. |
| **branch circuit—individual** | A circuit that supplies only one utilization equipment. |
| **breaker strip** | Thin strips of material placed between phase conductors and the grounding conductor in flat, parallel, portable cables; provides extra mechanical and electrical protection. |
| **breakout** | The point at which conductor(s) are taken out of a multiconductor assembly. |
| **British thermal unit (Btu)** | Quantity of heat required to raise the temperature of 1 pound of water 1° Fahrenheit. |
| **brush** | A conductor between the stationary and rotating parts of a generator, usually of carbon. |
| **Buna** | A synthetic rubber insulation. |
| **bus** | The conductor(s) serving as a common connection for two or more circuits. |
| **busbars** | The conductive bars used as the main current supplying elements of panelboards or switchboards. |

| | |
|---|---|
| **cable** | An assembly of two or more wires larger than no. 8 AWG. |
| **cable, aerial** | An assembly of one or more conductors and a supporting messenger. |
| **cable, armored** | A cable having a metal protection (armor). |
| **cable, belted** | A multiconductor cable having a layer of insulation over the assembled insulated conductors. |
| **cable clamp** | A device used to clamp around a cable to produce mechanical strain. |
| **cable, control** | Used to supply voltage (on and off). |
| **cable, duplex** | A twisted pair of cables. |
| **cable, parkway** | Designed for direct burial. |
| **cable, portable** | Used to transmit power to mobile equipment. |
| **cable, power** | Used to supply current (power). |
| **cable, service drop** | The cable from the utility line to the customer's property. |
| **cable, submarine** | Designed for crossing under navigable bodies of water. |
| **cable tray** | A rigid structure to support cables. |
| **caisson** | Sunken panel in a ceiling. |

| | |
|---|---|
| **candela (cd)** | Unit for luminous intensity. |
| **candlepower (cp)** | The unit of luminous intensity of a light source. In SI units, it is called *candela* (cd). |
| **capacitor** | It stores energy, blocks the flow of direct current, and permits the flow of alternating current to a degree depending on the capacitance and frequency. |
| **charge** | The quantity of positive and negative ions in or on an object (unit: coulomb). |
| **choke coil** | A coil used to limit the flow of alternating current while permitting direct current to pass. |
| **circuit** | A closed path through which current flows from a generator, through various components, and back to the generator. |
| **circuit breaker** | A device designed to open a circuit by nonautomatic means when it senses an overload. |
| **circular mil** | A unit for measuring the cross-sectional area of a conductor (used in MCM conductors). |
| **clearance** | The vertical space between a cable and its conduit. |
| **coated wire** | Wire given a thin coating of tin, lead, nickel, etc., for protection. |
| **code** | Short for National Electrical Code. |

| | |
|---|---|
| **coefficient of beam utilization (CBU)** | The ratio of the lumens reaching a specified area directly from a floodlight or spotlight to the total lumens emitted by the beam source. |
| **coefficient of utilization (CU)** | The ratio of the lumens from a luminaire received on a work plane to the lumens emitted by the luminaire's lamp(s). It depends on the particular space as well as on the fixture. |
| **coil** | A wire or cable wound in a series of closed loops. |
| **collector** | The part of a transistor that collects electrons. |
| **color code** | Identifying conductors by the use of color. |
| **color rendition** | The effect of a light source on the apparent color of objects in comparison with their apparent color under a reference light source. |
| **commutator** | Device used on electric motors or generators to maintain a unidirectional current. |
| **concealed** | Wires in concealed raceways are considered concealed, even though they may become accessible by withdrawing them. |
| **conductance** | The ability of material to carry an electric current. |
| **conductor** | Any substance that allows energy to flow through it; the wire used to carry electricity. |
| **conductor, bare** | Conductor with no insulation. |
| **conductor, insulated** | A conductor covered with insulation. |

| | |
|---|---|
| **conductor load** | The mechanical loads on an aerial conductor—wind, weight, ice, etc. |
| **conduit** | A channel or tube which carries conductors when they are in need of protection; a tubular raceway. |
| **conduit, rigid metal** | Conduit made of Schedule 40 pipe, normally 10-foot lengths. |
| **connector, pressure (solderless)** | A device which establishes the connection between two or more conductors or a terminal by means of mechanical pressure and without the use of solder. |
| **constant current** | A type of power system in which the same amount of current flows through each utilization equipment |
| **constant voltage** | The common type of power in which all loads are connected in parallel, and different currents flow through each load. |
| **continuous load** | Where the maximum current is expected to continue for three hours or more. |
| **continuous load, NEC** | In operation three hours or more. |
| **control** | Automatic or manual device used to stop, start, and/or regulate the flow of electricity. |
| **controller** | A device serving the electric power delivered to the apparatus to which it is connected. |
| **convenience outlet** | An outlet which receives the plugs of electrical devices such as lamps and radios. Also called *receptacle.* |

| | |
|---|---|
| **coulomb** | The derived SI unit for quantity of electricity or electrical charge (1 coulomb equals 1 ampere-second). |
| **copper** | In electricity, refers to copper conductors. |
| **cord** | A small, flexible conductor assembly. |
| **corona** | A low-energy electrical discharge caused by ionization of a gas by an electric field. |
| **current** | The time rate of flow of electric charges (unit: ampere). |
| **current density** | The current per unit cross-sectional area of a conductor. |
| **current, leakage** | That small amount of current which flows through insulation whenever a voltage is present and heats the insulation because of the insulation's resistance; the leakage current is in phase with the voltage, and is a power loss. |
| **current limiting** | Short-circuit protective devices, such as fuses and circuit breakers. |
| **cycle** | A set of operations that are repeated regularly in the same sequence. |
| **damp location** | A location subject to a moderate degree of moisture. |
| **dead-front** | A switchboard or panel or other electrical apparatus without "live" terminals. |
| **demand** | The measure of the maximum load of a utility's customer over a short period of time. |

**demand factor**

The ratio of the maximum demand of a system, or part of a system, to the total connected load of the system.

**demand load**

The actual amount of load on a circuit at any time.

**dielectric strength**

The maximum voltage which an insulation can withstand.

**diffuse reflection**

The reflected light is scattered in all directions so that the reflecting surface appears equally bright from any angle.

**diffuse transmission**

The transmission light is scattered evenly in all directions.

**direct current (dc)**

(1) Electricity which flows in only one direction; (2) the type of electricity produced by a battery and dc generators.

**direct transmission**

When light passes through clear, transparent materials. The angle at which the light leaves is the same as that at which it enters.

**disconnect**

A switch for disconnecting an electrical circuit or load.

**disconnecting means**

Means whereby the conductors of a circuit can be disconnected from their source of supply.

**displacement current**

The effective current flow across a capacitor.

**distortion**

Unfaithful reproduction of signals.

**diversity factor**

The ratio of the sum of load demands to a system demand.

**drawing**

Wiring diagram showing how devices are interconnected.

| | |
|---|---|
| **dry location** | A location not normally subject to dampness or wetness. |
| **dynamic** | A state in which one or more quantities exhibit appreciable change within an arbitrarily short time interval. |
| **dynamometer** | A device for measuring power output or power input of a mechanism. |
| **eddy currents** | Circulating currents induced in conducting materials by varying magnetic fields. |
| **Edison base** | The standard screw base used for ordinary lamps. |
| **efficacy** | Lumens per watt produced by a lamp. |
| **efficiency** | The ratio of the output to the input. |
| **efficiency, for lighting** | The efficiency of a luminaire is the ratio of light output (luminous flux) to the light produced by the lamp. |
| **elbow** | A short conduit which is bent. |
| **electric defrosting** | Use of electric resistance heating coils to melt ice and frost. |
| **electric heating** | House-heating system in which heat from electrical resistance units is used to heat rooms. |
| **electric water valve** | Solenoid-type (electrically operated) valve used to turn water flow on and off. |
| **electricity** | Relating to the flow or presence of charged particles; a fundamental physical force or energy. |

| | |
|---|---|
| **electrocution** | Death caused by electrical current through the heart. |
| **electrode** | A conductor through which current transfers to another material. |
| **electrolysis** | The production of chemical changes by the passage of current from an electrode to an electrolyte or vice versa. |
| **electrolyte** | A liquid or solid that conducts electricity by the flow of ions. |
| **electrolytic condenser-capacitor** | Plate or surface capable of storing small electrical charges. |
| **electromagnet** | A device consisting of a ferromagnetic core and a coil that produces appreciable magnetic effects only when an electric current exists in the coil. |
| **electromotive force (emf) voltage** | Electrical force that causes current (free electrons) to flow or move in an electrical circuit. Unit of measurement is the volt. |
| **electron** | The subatomic particle that carries the unit negative charge of electricity. |
| **electropneumatic** | An electrically controlled pneumatic device. |
| **electrotherapy** | The use of electricity in treatment of disease. |
| **electrothermics** | Direct transformation of electric and heat energy. |
| **emergency source** | Standby source of electric power; to be used when electric power is interrupted. |

| | |
|---|---|
| **enameled wire** | Wire insulated with a thin baked-on-varnish enamel. |
| **energy** | Kilowatt-hours (kWh) is equal to the product of power and time:<br><br>$$\text{Energy} = \text{power} \times \text{time}$$ |
| **engine** | An apparatus which converts heat to mechanical energy. |
| **equilibrium** | Properties are time constant. |
| **equipment** | A general term (including material, fittings, devices, appliances, fixtures, apparatus) used in connection with an electrical installation. |
| **excitation losses** | Losses in a transformer or electrical machine due to voltage. |
| **excite** | To initiate or develop a magnetic field. |
| **exposed** | (1) Refers to live parts that are not suitably guarded, isolated, or insulated; (2) in regard to wiring methods, *exposed* means not concealed. |
| **farad** | The basic unit of capacitance. |
| **feedback** | The process of transferring energy from the output circuit of a device back to its input. |
| **feeder** | Conductors in conduit or a busway run which carry a large block of power from the service equipment to a subfeeder panel. |
| **field, electrostatic** | The region near a charged object. |

**filament**            A cathode in the form of a metal wire in an electron tube.

**final**               The final inspection of an electrical installation.

**fire-shield cable**   Material or devices to prevent fire from spreading between raceways.

**flat wire**           A rectangular wire having 0.188 inch thickness or less, 1¼ inch width or less.

**footcandle (fc)**     Unit of light flux density, equal to 1 lumen per square foot.

**footlamberts (FL)**   (1) A quantitative unit for measuring brightness; (2) brightness, or luminance, is an index of the intensity of light being emitted, transmitted, or reflected from a surface.

**fossil fuels**        Oil, gas, and coal.

**frequency**           The number of complete cycles an alternating electric current, sound wave, or vibrating object undergoes per second.

**fuse**                A protecting device which opens a circuit due to overcurrent passing through.

**fuse, nonrenewable**  A fuse which must be replaced after it interrupts a circuit.

**fusible plug**        A plug or fitting made with a metal of a known low melting temperature; used as a safety device in case of fire.

**galvanometer**        Used for measuring a small electrical current by means of a mechanical motion derived from electromagnetic or dynamic forces.

| | |
|---|---|
| **ganged switches** | A group of switches arranged next to each other in ganged outlet boxes. |
| **gem box** | The most common rectangular outlet box used to hold wall switches and receptacle outlets. |
| **general-use switch** | A switch intended for use in general distribution and branch circuits. |
| **general-use snap switch** | A switch that can be installed in flush device boxes. |
| **generator** | A rotating machine used to convert mechanical energy to electrical energy. |
| **glare** | The effect of excessive brightness in the field of view. It may be direct from a light source, or reflected from a shiny surface. |
| **ground** | A large conducting body (such as the earth) used as a common return for an electric circuit and as an arbitrary zero of potential. |
| **grounded** | Connected to earth. |
| **grounded conductor** | A system or circuit conductor that is intentionally grounded. |
| **grounding** | The device or conductor connected to ground designed to conduct in abnormal conditions only. |
| **ground-fault interrupter (gfi)** | A protective device that detects abnormal current flowing to ground and then interrupts the circuit. |
| **half effect** | The changing of current density in a conductor due to a magnetic field extraneous to the conductor. |

| | |
|---|---|
| **half wave** | Rectifying only half of a sinusoidal ac supply. |
| **henry** | The derived SI unit for inductance (1 henry equals 1 weber per ampere). |
| **hertz** | The derived SI unit for frequency (1 hertz equals one cycle per second). |
| **hickey** | (1) A conduit-bending tool; (2) a box fitting for hanging lighting fixtures. |
| **home run** | That part of a branch circuit from the panelboard housing the branch circuit fuse. |
| **horsepower** | The non-SI unit power: 1 hp = 1 HP = 746 W (electric) = 9800 W (boiler). |
| **hot** | Energized with electricity. |
| **hot leg** | A circuit conductor which normally operates at a voltage above ground; the phase wires or energized circuit. |
| **ignition transformer** | A transformer designed to provide a high-voltage current. |
| **impedance ($Z$)** | The opposition to current flow in an ac circuit; impedance includes resistance $R$, capacitive reactance $XC$, and inductive reactance $XL$ (unit: ohm). |
| **impedance matching** | Matching source and load impedance for optimum energy transfer with minimum distortion. |

**impulse**                        A surge of unidirectional polarity.

**inductance**                     A voltage due to a time-varying current; the opposition to current change, causing current changes to lag behind voltage changes (unit: henry).

**individual branch circuit**      One that supplies only a single piece of electrical equipment.

**inductor**                       A device having winding(s) with or without a magnetic core for creating inductance in a circuit.

**infrared lamp**                  An electrical device that emits infrared rays—invisible rays just beyond red in the visible spectrum.

**infrared radiation**             Radiant energy given off by heated bodies; transmits heat and will pass through glass.

**ink**                            The material used for legends and color coding.

**in phase**                       The condition existing when waves pass through their maximum and minimum values of like polarity at the same instant.

**insulation, electrical**         A medium in which it is possible to maintain an electrical field with little supply of energy from additional sources.

**insulation, class rating**       A temperature rating descriptive of classes of insulations.

**inverter**                       An item which changes dc to ac.

**jack**                           A plug-in type terminal.

**jacket**

A nonmetallic, polymeric, close-fitting protective covering over cable insulation; the cable may have one or more conductors.

**jacket**

Conducting: an electrically conducting polymeric covering over an insulation.

**joule**

The derived SI unit for energy, work, quantity of heat: 1 joule equals 1 newton-meter.

**jumper**

A short length of conductor, usually a temporary connection.

**junction**

A connection of two or more conductors.

**junction box**

Group of electrical terminals housed in a protective box or container.

**kelvin (K)**

The basic SI unit of temperature: $1/273.16$ of thermodynamic temperature of the triple point of water.

**kilowatt**

Unit of electrical power equal to 1000 watts.

**kilowatt-ft**

The product of load in kilowatts and the circuit's distance over which a load is carried, in feet; used to compute voltage drop.

**kinetic energy**

Energy by virtue of motion.

**Kirchoff's laws**

The algebraic sum of the currents at any point in a circuit is zero. The algebraic sum of the product of the current and the impedance in each conductor in a circuit is equal to the electromotive force in the circuit.

**knockout**

A portion of an enclosure designed to be readily removed for installation of a raceway.

| | |
|---|---|
| **lamp** | A device to convert electrical energy to radiant energy, normally visible light; only 10 to 20 percent of electric energy is converted to light. Incandescent filament lamps are commonly referred to as *bulbs,* fluorescent lamps are called *tubes,* and H.I.D. light sources are simply called *lamps.* |
| **law of charges** | Like charges repel; unlike charges attract. |
| **law of magnetism** | Like poles repel; unlike poles attract. |
| **leg** | A portion of a circuit. |
| **Lenz's law** | "In all cases the induced current is in such a direction as to oppose the motion which generates it." |
| **lighting outlet** | An outlet intended for the direct connection of a lamp holder and lighting fixture. |
| **lighting system** | A method of describing in which direction the light is emitted from the fixture. Systems are direct, semidirect, direct-indirect, general diffuse, semi-indirect, and indirect. |
| **lightning arrestor** | A device designed to protect circuits and apparatus from high transient voltage by diverting the overvoltage to ground. |
| **line-voltage thermostat** | A thermostat that is connected directly to the line. Full power is fed through it to the controlled heater or air conditioner. |
| **live** | Energized. |

| | |
|---|---|
| **live-front** | Any panel or other switching which has exposed electrically energized parts on its front, presenting the possibility of contact by personnel. |
| **load center** | An assembly of circuit breakers or switches. |
| **load factor** | The ratio of the average to the peak load over a period. |
| **load losses** | Those losses incidental to providing power. |
| **load side** | The side of a device electrically farthest from the current source. |
| **low-voltage switching** | A system of outlet control by low-voltage switches and relays. |
| **lumen (lm)** | A quantity of light used to express the total output of a light source. It is a unit of measure for luminous flux. One lumen is the amount of light which falls on 1 sq ft of surface area (when all points on the surface are 1 ft from a standard candle). |
| **luminaire** | A complete lighting unit, consisting of a lamp or lamps, designed to distribute the light. A luminaire is commonly referred to as a *fixture*. |
| **luminous flux** | The radiant energy which the human eye can see; it produces the sensation of light in the human eye. |
| **lus** | The derived SI unit for illuminance. |
| **lux** | Metric unit of lighting flux density or illumination. |
| **machine** | An item to transmit and modify force or motion, normally to do work. |

| | |
|---|---|
| **magnet** | A body that produces a magnetic field external to itself; magnets attract iron particles. |
| **magnetic field** | A magnetic field in alternating current is the force field established by ac through a conductor, especially a coiled conductor. |
| **magnetic pole** | Those portions of the magnet toward which the external magnetic induction appears to converge (south) or diverge (north). |
| **master (central) control** | Control of all the outlets from one point. |
| **matte surface** | A surface from which reflection is predominantly diffused. |
| **megohmmeter** | An instrument for measuring extremely high resistance. |
| **messenger** | The supporting member of an aerial cable. |
| **metal clad (MC)** | The cable core is enclosed in a flexible metal covering. |
| **metal-clad switchgear** | Switchgear having each power circuit device in its own metal-enclosed compartment. |
| **meter (electrical meter)** | Used to measure energy = power $\times$ time. |
| **meter pan** | A metal enclosure with a round opening through which a kilowatt-hour meter is mounted. |
| **microwave** | Radio waves of frequencies above one gigahertz. |
| **molded-case breaker** | A circuit breaker enclosed in an insulating housing. |

**molecule**　　　The group of atoms which constitutes the smallest particle in which a compound or material can exist separately.

**motor**　　　An apparatus to convert from electrical to mechanical energy.

**motor, capacitor**　　　A single-phase induction motor with an auxiliary starting winding connected in series with a condenser for better starting characteristics.

**motor circuit switch**　　　A switch, rated in horsepower, capable of interrupting the maximum operating overload current of a motor.

**motor control**　　　Device to start and/or stop a motor at certain temperature or pressure conditions.

**motor control center**　　　A grouping of motor controls such as starters.

**negative**　　　Connected to the negative terminal of a power supply.

**network**　　　An aggregation of interconnected conductors consisting of feeders, mains, and services.

**network limiter**　　　A current-limiting fuse for protecting a single conductor.

**neutral**　　　The element of a circuit from which other voltages are referenced with respect to magnitude and time displacement in steady-state conditions.

**neutral block**　　　The neutral terminal block in a panelboard, meter enclosure, gutter, or other enclosure in which circuit conductors are terminated or subdivided.

**neutral wire**     A circuit conductor which is common to the other conductors of the circuit, having the same voltage between it and each of the other circuit wires and usually operating grounded (such as the neutral of three-wire single-phase, or three-phase four-wire wye systems).

**neutron**     Subatomic particle contained in the nucleus of an atom: electrical neutral.

**ohm**     The derived SI unit for electrical resistance or impedance: 1 ohm equals 1 volt per ampere.

**ohmmeter**     An instrument for measuring resistance in ohms.

**Ohm's law**     The relationship between current and voltage in a circuit. It states that current is proportional to voltage and inversely proportional to resistance. Expressed algebraically: in dc circuits $I = V/R$; in ac circuits $I = V/Z$.

**open**     A circuit which is energized by not allowing useful current to flow.

**opening time**     The period in which an activation signal is initiated and switch contacts part.

**oscillation**     The variation, usually with time, of the magnitude of a quantity which is alternately greater and smaller than a reference.

**oscillator**     A device that produces an alternating or pulsating current or voltage electronically.

**outlet**     A point on a wiring system at which electric current is taken off to supply an electric load.

| | |
|---|---|
| **outlet box** | A metal box containing wires from a branch circuit connected to wires from electric load. |
| **outline lighting** | An arrangement of incandescent lamps or gaseous tubes. |
| **output** | The energy delivered by a circuit or device; the terminals for such delivery. |
| **overcurrent device** | A device such as a fuse or a circuit breaker designed to protect a circuit against excessive current by opening the circuit. |
| **overload** | A condition of excess current (more current flowing than the circuit was designed to carry). |
| **overvoltage (cable)** | Voltages above normal operating voltages. |
| **pad-mounted** | A *pad-mount transformer* is a completely enclosed transformer mounted outdoors on a concrete pad. |
| **pan** | A sheet-metal enclosure for a watt-hour meter, called a *meter pan*. |
| **panel** | A unit for one or more sections of flat material suitable for mounting electrical devices. |
| **panelboard** | A single panel or group of panel units designed for assembly in the form of a single panel. |
| **panel directory** | A listing of the panel circuits appearing on the panel door. |
| **panel schedule** | A schedule appearing on the electrical drawings detailing the equipment contained in the panel. |

| | |
|---|---|
| **parallel** | Connections of two or more devices between the same two terminals of a circuit. |
| **parallel circuit** | One where all the elements are connected across the voltage source. |
| **payoff** | The equipment to guide the feeding of wire. |
| **phase angle** | The measure of the progression of a periodic wave in time or space from a chosen instant or position. |
| **phase conductor** | The conductors other than the neutral. |
| **phase leg** | One of the phase conductors (an ungrounded or "hot" conductor) of a polyphase electrical system. |
| **phaseout** | A procedure by which the individual phases of a polyphase circuit or system are identified. |
| **photocell** | A device in which the current-voltage characteristic is a function of incident radiation (light). |
| **photoelectric control** | A control sensitive to incident light. |
| **photoelectricity** | A physical action wherein an electrical flow is generated by light waves. |
| **PILC cable** | Paper insulated, lead covered. |
| **pilot lamp** | A lamp that indicates the condition of an associated circuit. |

| | |
|---|---|
| **pilot wire** | An auxiliary insulated conductor in a power cable used for control or data. |
| **plug** | A male connector for insertion into an outlet. |
| **plug-in bus duct** | Bus duct with built-in power tap-off points. Tap-off is made with a plug-in switch or circuit breaker. |
| **polarity** | The directions of current flow in a dc circuit. By convention, current flows from + to −. Electron flow is actually in the opposite direction. |
| **pole** | That portion of a device associated exclusively with one electrically separated conducting path of the main circuit of device; a supporting circular column. |
| **potential** | The difference in voltage between two points of a circuit. |
| **potentiometer** | An instrument for measuring an unknown voltage. |
| **pothead** | A terminator for high-voltage circuit conductor to keep moisture out of the insulation and to protect the cable end. |
| **power (P)** | Expressed in watts (w) or kilowatts (kW):<br>In dc circuit, $P = VI$ and $P = I^2R$<br>In ac circuit, $P = VI \times$ power factor |
| **power factor (PF)** | A quantity that relates the volt-amperes of an ac circuit to the wattage or power:<br>Power = volt-amperes × power factor |
| **power loss (cable)** | Losses due to internal cable impedance, mainly $I^2R$; the losses cause heating. |

**primary source**  A luminous source where light energy is generated and transmitted directly to a task.

**protector, circuit**  An electrical device that will open an electrical circuit if excessive electrical conditions occur.

**proton**  The hydrogen atom nucleus; it is electrically positive.

**quadruplexed**  Twisting of four conductors together.

**qualified person**  A person familiar with construction, operation, and hazards.

**raceway**  Any support system, open or closed, for carrying electric wires.

**rack (cable)**  A device to support cables.

**radiant energy**  Energy traveling in the form of electromagnetic waves.

**reactance**  The imaginary part of impedance. The opposition to ac due to capacitance ($XC$) and inductance ($XL$).

**radiant cables**  Electric cables embedded in the ceiling for heating.

**rating, temperature (cable)**  The highest conductor temperature attained in any part of the circuit during normal operation, emergency overload, or short circuit.

**receptacle wires**  Number of connecting wires including the ground wire.

**recessed**  A convector cabinet that extends partially or fully into the wall.

| | |
|---|---|
| **reflector** | A device for redirecting the radiant energy of a lamp by reflecting it in a desired direction. |
| **refraction** | The bending of a ray of light as it passes obliquely through a material. |
| **refractor** | A device for redirecting the radiant energy of a lamp in the desired direction by refraction through a lense. |
| **reflection factor or**<br>   **reflectance** | Ratio between light reflected from, and light falling on, an object. |
| **remote control (RC) switch** | A magnetically operated mechanically held switch, normally used for remote switching of blocks of power. |
| **resistance (R)** | The unit in an electric circuit analogous to friction in a hydraulic circuit (expressed in ohms). |
| **resonance** | In a circuit containing both inductance and capacitance. |
| **Romex** | One of several trade names for NEC-type NM nonmetallic sheathed flexible cable. |
| **safety motor control** | Electrical device used to open a circuit if the temperature, pressure, and/or the current flow exceed safe conditions. |
| **safety plug** | Device that will release the contents of a container above normal pressure conditions and before rupture pressures are reached. |
| **secondary** | The second circuit of a device or equipment, which is not normally connected to the supply circuit. |

| | |
|---|---|
| **self-inductance** | Magnetic field induced in the conductor carrying the current. |
| **semiconductor** | A material that has electrical properties of current flow between a conductor and an insulator. |
| **series circuit** | One with all the elements connected end to end. The current is the same throughout, but the voltage can be different across each element. |
| **service** | The equipment used to connect to the conductors run from the utility line, including metering, switching, and protective devices. |
| **service conductors** | The supply conductors that extend from the street main or transformers to the service equipment of the premises being supplied. |
| **service drop** | Run of cables from the power company's aerial power lines to the point of connection to a customer's premises. |
| **service entrance** | The point at which power is supplied to a building. |
| **service equipment** | Consisting of a circuit breaker or switch and fuses and accessories. |
| **service lateral** | The underground service conductors between the street main or transformers and the service-entrance conductors in a terminal box or meter. |
| **service raceway** | The rigid metal conduit; electrical metallic tubing. |
| **service switch** | One to six disconnect switches or circuit breakers. |

**sidewall load**      The normal force exerted on a cable under tension at a bend; quite often called *sidewall pressure.*

**signal**      A detectable physical quantity or impulse (such as a voltage, current, or magnetic field strength) by which messages or information can be transmitted.

**signal circuit**      Any electrical circuit supplying energy to an appliance that gives a recognizable signal.

**single-phase circuit**      An ac circuit having one source voltage supplied over two conductors.

**single-phase motor**      Electric motor that operates on single-phase alternating current.

**single pole**      Connects to a single hot line.

**6-foot rule**      The NEC rule that no point along a wall area be more than 6 ft from a wall outlet.

**shop drawings**      Contractor's or manufacturer's drawings giving equipment construction details.

**short circuit**      An electric circuit with zero load; an electrical fault.

**solar cell**      The direct conversion of electromagnetic radiation into electricity; certain combinations of transparent conducting films separated by thin layers of semiconducting materials.

**solid-type PI cable**      A pressure cable without constant pressure controls.

| | |
|---|---|
| **specular reflection** | Mirrorlike reflection. |
| **splice** | The electrical and mechanical connection between two pieces of cable. |
| **split-phase motor** | Motor with two stator windings. Winding in use while starting is disconnected by a centrifugal switch after the motor attains speed; then the motor operates on the other winding. |
| **squirrel-cage motor** | An induction motor having the primary winding (usually the stator) connected to the power and a current is induced in the secondary cage winding (usually the rotor). |
| **starting relay** | An electrical device that connects and/or disconnects the starting winding of an electric motor. |
| **starting winding** | Winding in an electric motor used only during the brief period when the motor is starting. |
| **superconductors** | Materials whose resistance and magnetic permeability are infinitesimal at absolute zero (−273°C). |
| **supervised circuit** | A closed circuit having a current-responsive device to indicate a break or ground. |
| **surge** | A sudden increase in voltage and current; a transient condition. |
| **switch** | A device for opening and closing or for changing the connection of a circuit. |
| **switch, ac general-use snap** | A general-use snap switch suitable only for use on alternating current circuits. |

| | |
|---|---|
| **switch, ac-dc general-use snap** | A type of general-use snap switch suitable for use on either direct- or alternating-current circuits. |
| **switch, general-use** | A switch intended for use in general distribution and branch circuits. It is rated in amperes. |
| **switch, isolating** | A switch intended for isolating an electrical circuit from the source of power. |
| **switch, knife** | A switch in which the circuit is closed by a moving blade engaging contact clips. |
| **switch-leg** | That part of a circuit run from a lighting outlet down to an outlet box which contains the wall switch. |
| **switch, motor-circuit** | A switch, rated in horsepower, capable of interrupting the maximum operating overload current of a motor having the same horsepower rating as the switch at the rated voltage. |
| **system voltage** | Voltage from the power company; transformer voltage. |
| **testlight** | Light provided with test leads that is used to test or probe electrical circuits to determine if they are energized. |
| **test, voltage-breakdown** | Method of applying a multiple of rated voltage to a cable for several minutes, then increasing the applied voltage by 20 percent for the same period until breakdown; applying a voltage at a specified rate until breakdown. |
| **test, voltage-life** | Applying a multiple of rated voltages over a long time period until breakdown. Time to failure is the parameter measured. |

| | |
|---|---|
| **test, volume-resistivity** | Measuring the resistance of a material such as the conducting jacket or conductor stress control. |
| **three-phase circuit** | A polyphase circuit of three interrelated voltages for which the phase difference is 120°; the common form of generated power. |
| **three-way switch** | A three-terminal switch that connects $X$-$Y$ or $X$-$Z$, where $X$ is common. |
| **three-way switching** | An arrangement for controlling (an) outlet(s) from two locations. |
| **timers** | Mechanism used to control on and off times of an electrical circuit. |
| **transfer switch** | A device for transferring one or more load-conductor connections from one power source to another. |
| **transformer** | A static device consisting of winding(s), with or without a tap(s), with or without a magnetic core, for introducing mutual coupling by induction between circuits. |
| **transformer, potential** | Designed for use in measuring high voltage. Normally, the secondary voltage is 120 V. |
| **transformer, power** | Designed to transfer electrical power from the primary circuit to the secondary circuit(s) to (1) step up the secondary voltage at less current or (2) step down the secondary voltage at more current. The voltage-current product is constant for either primary or secondary. |
| **transformer, rectifier** | Combination transformer and rectifier in which input in ac may be varied and then rectified into dc. |

**transformer, safety isolation**  Inserted to provide a nongrounded power supply such that a grounded person accidentally coming in contact with the secondary circuit will not be electrocuted.

**transformer, vault-type**  Suitable for an operation occasionally submerged in water.

**transistor**  An active semiconductor device with three or more terminals.

**transmission**  Transfer of electric energy from one location to another through conductors or by radiation or induction fields.

**transmission line**  A long electrical circuit.

**transposition**  Interchanging the position of conductors to neutralize interference.

**triode**  A three-electrode electron tube containing an anode, a cathode, and a control electrode.

**triplex**  Three cables twisted together.

**trolley wire**  Solid conductor designed to resist wear due to rolling- or sliding-current pickup trolleys.

**trunk feeder**  A feeder connecting two generating stations or a generating station and an important substation.

**two-phase**  A polyphase ac circuit having two interrelated voltages.

**ultrasonic**  Sounds having frequencies higher than 20 KHz; 20 KHz is at the upper limit of human hearing.

| | |
|---|---|
| **undercurrent** | Lower-than-normal operating current. |
| **undervoltage** | Lower-than-normal operating voltage. |
| **ungrounded** | Intentionally not connected to the earth. |
| **universal motor** | A motor designed to operate on either ac or dc at about the same speed and output. |
| **utilization equipment** | Equipment that uses electric energy for mechanical, chemical, heating, lighting, etc. |
| **utilization voltage** | The voltage that is utilized; motor voltage. |
| **voltage (*V*)** | The electric pressure in an electric circuit expressed in volts; the electrical property that provides the energy for current flow. |
| **voltage drop** | The voltage drop around a circuit including wiring and loads must equal the supply voltage. |
| **voltage, breakdown** | The minimum voltage required to break down an insulation's resistance, allowing a current flow through the insulation, normally at a point. |
| **voltage rating of a cable** | Phase-to-phase ac voltage when energized by a balanced three-phase circuit having a solidly grounded neutral. |
| **voltage regulator** | A device to decrease voltage fluctuations to loads. |
| **voltage, signal** | Voltages to 50 V. |

**voltage to ground**  The voltage between an energized conductor and earth.

**voltage, UHV**

**(ultrahigh voltage)**  765 + kV.

**voltmeter**  An instrument for measuring voltage.

**watt**  The derived SI unit for power, radiant flux: 1 watt equals 1 joule per second.

**watt-hour**  The number of watts used in one hour.

**watt-hour meter**  A meter which measures and registers the integral, with respect to time, of the active power in a circuit. 1000 Wh = 1 kWh.

**wavelength**  The distance measured along the direction of propagation between two points which are in phase on adjacent waves.

**wet-cell battery**  A battery having a liquid electrolyte.

**winding**  An assembly of coils designed to act in consort to produce a magnetic flux field or to link a flux field.

**wire**  A slender rod or filament of drawn metal.

**wire bar**  A cast shape which has a square cross section with tapered ends.

**wire, building**  A class of wire and cable, usually rated at 600 V (used for interiors).

**wire, hookup**  Insulated wire for low voltage and current.

**wire, resistance**	Wire having appreciable resistance.

**wireway**	Term generally used to mean a surface raceway.

**wiring diagram**	Diagram showing actual wiring, with numbered terminals.

**wiring device**	Receptacle, switch, or pilot light that is wired in a branch circuit.

# CONTENTS

## Part 2

# ELECTRICITY

E

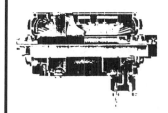

# Generating Electricity

# Electric Systems Components and Calculations

# ELECTRICITY

## E-1 ELECTRIC ENERGY

William Gilbert, **father of modern electricity,** in 1600 A.D. described the **"electrification of many matters."**

There are 105 different matters known as of the present date.


## E-2 ELECTRICITY PRODUCED BY NATURE

The natural forms of electricity include:

1. *Electromagnetic radiation* **from the sun.** Several forms of solar energy have been harnessed for practical use.

2. *Lightning,* which causes damage to structures.

3. *Galvanic cells,* which cause corrosion in the pipes, etc.

4. *Static discharge.* Electric charge at rest.


## E-3 ELECTRICITY AND ELECTRONS

Electric phenomena are associated by electrons, one of the smallest components of matter.

The basic block of matter consists of:

1. *Electrons,* which are negatively charged

2. *Neutrons* (free electrons), which are negatively charged

3. *Protons,* which are positively charged

Atoms of any matter are a combination of electrons, free electrons, and protons.

In a normal atom, the charge of the electrons and protons balance, so that when a free electron leaves an atom, the atom becomes positively (+) charged; if a free electron is added, the atom becomes negatively (−) charged.

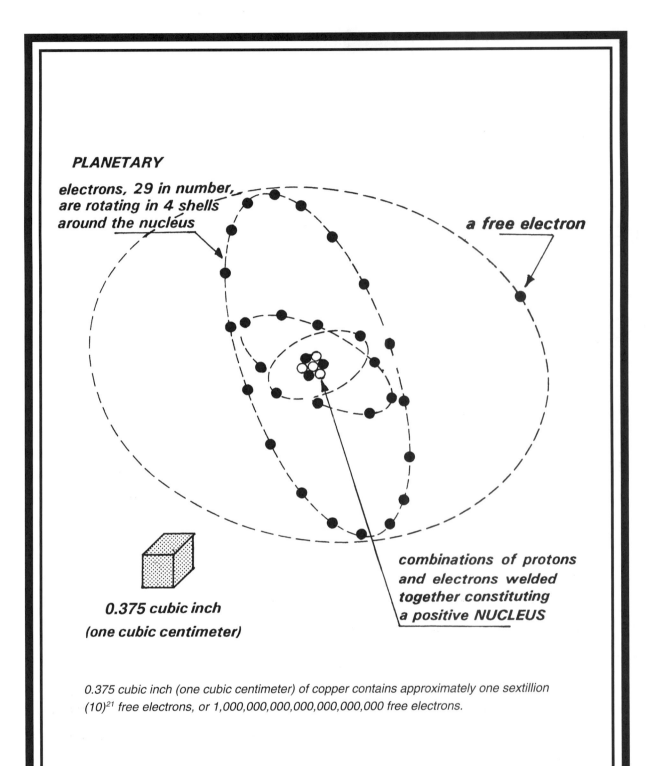

PLANETARY electrons, 29 in number, are rotating in 4 shells around the nucleus

a free electron

combinations of protons and electrons welded together constituting a positive NUCLEUS

0.375 cubic inch
(one cubic centimeter)

0.375 cubic inch (one cubic centimeter) of copper contains approximately one sextillion $(10)^{21}$ free electrons, or 1,000,000,000,000,000,000,000 free electrons.

**Figure E-1   Atom of Copper**

## E-4    ELECTRICITY MOVING IN CONDUCTORS

The best conductors of current are the metals which contain the highest number of free electrons, such as platinum, gold, silver, copper, and aluminum.

Copper and aluminum are used as conductors in electrical systems.

Aluminum has approximately 80 percent of the conductivity of copper.

The atoms of copper have various combinations of protons and electrons welded together, constituting a **positive nucleus.**

**Planetary electrons,** 29 in number, rotate in four shells around the nucleus (Fig. E-1).

The outmost planetary electron (**free electron**) is weakly attracted to the nucleus and moves from one atom to another atom.

The free electrons constitute a small percentage of the total number of electrons in the copper; nevertheless, they are numerous.

0.375 cubic inch (one cubic centimeter) of copper contains approximately one sextillion $10^{21}$ free electrons, or 1,000,000,000,000,000,000,000 free electrons.

The motion of these free electrons in a solid conductor causes the current to move at the speed of light, 186,000 miles per second.

*Photo from Ewing Galloway*

*An old water mill for grinding corn.*

*Courtesy of Travel Magazine*

*This Japanese laborer is forcing water into an irrigation ditch by making the wheel go around. What crops need irrigation in Japan?*

**Figure E-2   Water Wheels**

# PRODUCTION OF ELECTRICITY

E-5    **HOW ELECTRICITY IS PRODUCED**

To the present date, the following remain undetermined:

1.    **Exactly what is electricity?**

2.    **How does it work?**

However, we know the following:

1.    **How to produce electricity**

2.    **How to use it**

The prime movers **turbine** and **generator** are the basic components of the **generation of electricity**.

Turbine is a rotary engine that changes the **force** of water, fuel, and wind into **mechanical energy** capable of rotating a shaft which is connected to a generator.

A generator (E-20 and E-24) is a machine used to change the mechanical energy into **electrical energy.**

E-6    **TURBINES PRODUCING MECHANICAL ENERGY**

There are three types of turbines used in electric power plants:

1.    *Water-driven turbines* **(E-7 and E-8)**

2.    *Steam-driven turbines* **(E-9)**

3.    *Nuclear reactor turbines* **(E-15)**

There are three other methods used to produce electrical energy within a project:

1.    *Internal-combustion engine* **(E-16)**

2.    *Wind turbine* **(E-17 and E-18)**

3.    *Solar photovoltaic cells* **(E-19)**

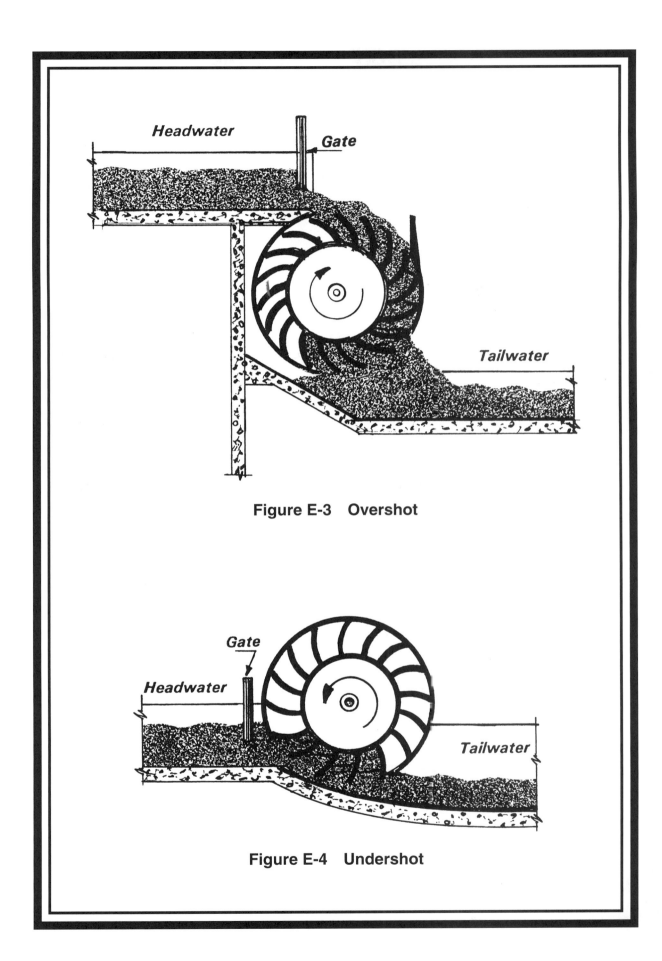

**Figure E-3  Overshot**

**Figure E-4  Undershot**

## E-7    WATER-DRIVEN TURBINES

The principle of the water-driven turbine has been known since ancient times.

In early 1880 water turbines were designed with fixed guide vanes, carved blades, or buckets to rotate a shaft connected to wheels used to grind the grain or sawing wood, etc. (Fig. E-2). They are still in operation in the United States and other parts of the world.

There are two types of water turbines:

1.   **Impulse type**

     a.   Activated by the force of water falling on the curved blades, called ***overshot*** (Fig. E-3)

     b.   Activated by the force of water running under the curved blades, called ***undershot*** (Fig. E-4)

2.   **Reaction type**

     The kinetic energy of the flow of water from nozzles on the bladed wheel provides the smooth continuity of rotation which is needed to operate the generator. The reaction type is commonly used in the hydroelectric power plants (Fig. E-5).

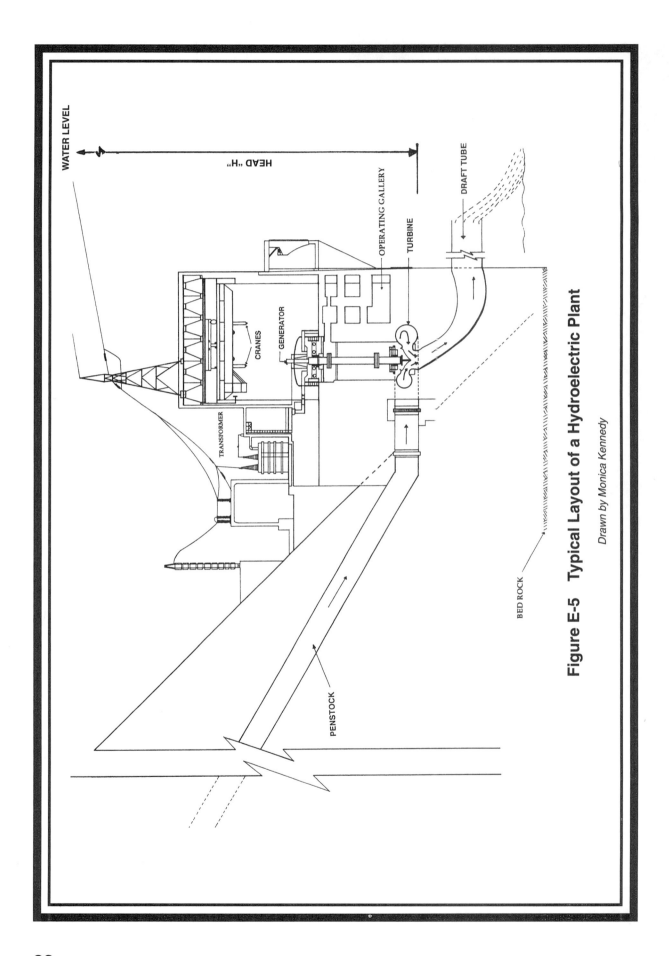

**Figure E-5  Typical Layout of a Hydroelectric Plant**

*Drawn by Monica Kennedy*

## E-8  HYDROELECTRIC POWER PLANTS

The water wheel has given way to the water turbine connected by a shaft to electric generators for the production of hydroelectric power. In the United States many enormous hydroelectric power plants are contributing substantially to the nation's electrical power supply. These plants include those at the Central Valley Project, Hoover Dam, Bonneville Dam (Fig. E-7), Niagara Falls Project, Grand Coulee Dam, and Parker Dam (Fig. E-6).

The kilowatts obtained from water power is as follows:

$$kW = \frac{Q \times H}{709 \text{ kW/cfm ft}} \times E_{TU} \times E_{TR} \times E_G$$

where  $kW$ = kilowatt

$Q$ = flow of water in cfm (cubic feet per minute)

$H$ = head (vertical height of source of water) in feet

$E_{TU}$ = efficiency of turbine

$E_{TR}$ = efficiency of transmission

$E_G$ = efficiency of generator

### *Example*

The flow of water is 100,000 cfm and the net head of water is 150 feet. If

$$E_{TU} = 85\%, \qquad E_{TR} = 95\%, \qquad \text{and} \qquad E_G = 85\%$$

what is the total kW produced?

### *Solution*

$$kW = \frac{100,000 \text{ cfm} \times 150 \text{ ft}}{709 \text{ kW/cfm ft}} = 21,156.56 \text{ kW}$$

$$21,156.56 \text{ kW} \times 0.85 \times 0.95 \times 0.85 = 14,521.34 \text{ kW}$$

**Figure E-6    Parker Dam Power Plant on the Colorado River**

**Figure E-7    Bonneville Dam on the Columbia River**

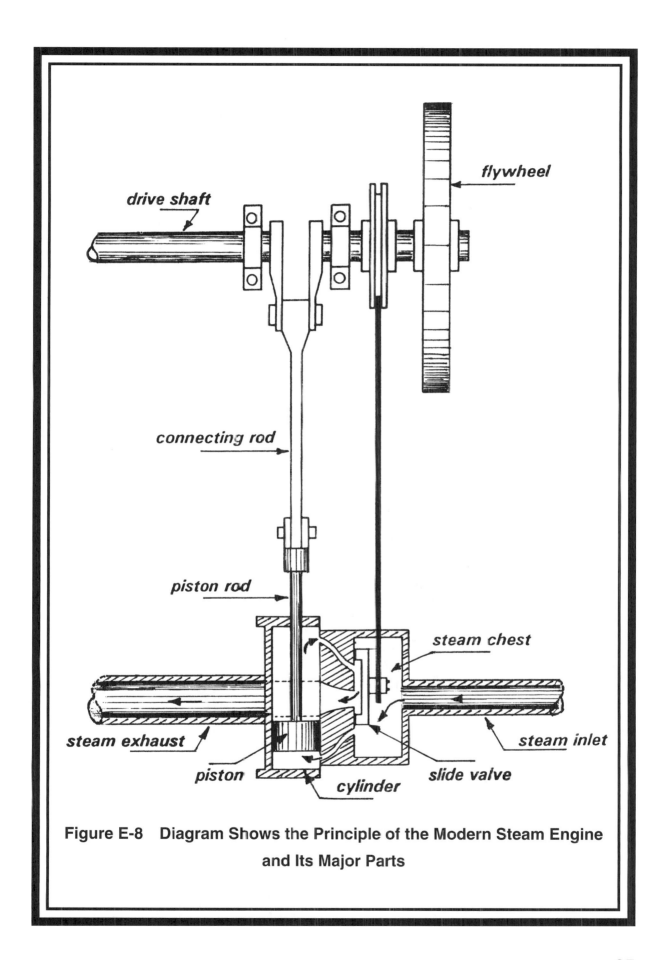

**Figure E-8 Diagram Shows the Principle of the Modern Steam Engine and Its Major Parts**

**Turbine Generator**

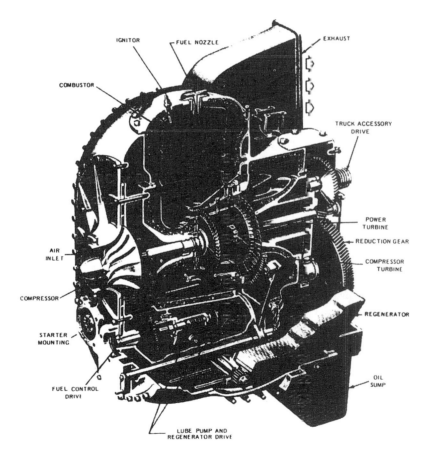

IGNITOR — FUEL NOZZLE — EXHAUST

COMBUSTOR

TRUCK ACCESSORY DRIVE

POWER TURBINE

REDUCTION GEAR

COMPRESSOR TURBINE

AIR INLET

COMPRESSOR

REGENERATOR

STARTER MOUNTING

OIL SUMP

FUEL CONTROL DRIVE

LUBE PUMP AND REGENERATOR DRIVE

**Figure E-9   Section Through Low-Pressure Gas Turbine Engine**

THE ILLUSTRATED COLUMBIA ENCYCLOPEDIA

## E-9    STEAM-DRIVEN TURBINES

A steam engine is a machine for converting *heat energy* into *mechanical energy* through the medium of *steam.*

The volume of water increases approximately 1600 times when it is converted into steam. The force obtained by this conversion is the basis of all steam turbines.

The practical solution to harnessing steam came about when James Watt patented his steam engine in 1769 (please see **James Watt,** G-12).

The modern version of his discovery is shown in Fig. E-8.

Watt's discoveries prepared the steam engine for the start of the **Industrial Revolution.**

In 1888 **Charles A. Parsons** of Great Britain and **C. G. P. de Laval** of Sweden were pioneers in the building of *steam turbines.*

In the 20th century, steam turbines became the prime movers of the largest parts of all electric generation equipment.

The basic principle of the steam turbine is that the jets of steam produced are directed through a stage or series of curved vanes on a rotating wheel.

The steam leaving a revolving wheel is directed onto second, third, and so on revolving wheels, causing the shaft to rotate the generator (Fig. E-9).

A drilling crew at work in an American coal mine. Holes are drilled in the coal "face," into which explosives are inserted, and chunks of the fuel are blasted from the bed.

The formation of bituminous coal began some 250 million years ago in great swamps, which are believed to have resembled this illustration.

**Figure E-10    Formation and Mining the Coal**

# FORCE OF FUELS

### E-10   FUELS USED TO PRODUCE STEAM

Fuel is a material that when burned in air produces **heat.**

The hydrogen and carbon in fuel rapidly combine with oxygen in the air **exothermal reaction** and create heat.

The fuels used to produce steam for turbines are:

a.   *Coal*

b.   *Fuel oil*

c.   *Natural gas*

d.   *Nuclear*

### E-11   COAL

Coal is composed of carbon and mineral matter. Coal originated in swampy fields with plant material some 250 million years ago in **the carboniferous period** of geologic time. Coal was formed under great pressure. The greater the stress extended in the process of **metamorphism,** the higher was the grade of coal produced.

There are many types of coal, starting with **peat coal** (lowest in carbon content) to **anthracite coal** (almost pure carbon) (Fig. E-10). **Bituminous coal** is high grade in carbon content and does not crumble easily. It is a deep black color, burns readily, and is used extensively as fuel in industries and power plants. The use of coal in electric power plants is restricted because of air pollution.

**Bituminous coal produces 14,600 Btu per pound, with an efficiency of 70 percent.**

### E-12   FUEL OIL

Fuel oil is one of the products obtained from **crude petroleum** (crude oils).

Crude petroleum is an oily, inflammable liquid, usually a greenish blue or dark brown in color. At ordinary temperature it is lighter than water. It is a mixture of hydrogen and carbon. **Hydrocarbons** also contain nitrogen, sulphur, and oxygen.

*One of the first photos of E. L. Drake's well, drilled in 1859 at Oil Creek, later Titusville, Pa.*

*Cracking units of Baytown, Texas, refineries.*

## Figure E-11   Production of Petroleum Products

Crude petroleum is refined by a process called *fractional distillation.*

Different components of petroleum have different boiling points. By boiling the crude oil at different temperature ranges, the portions of oil given off by this process are called *fraction.*

Components from different crude petroleum obtained are petroleum, lubricating oil, paraffin, *fuel oil,* and mineral oils (Fig. E-11).

Fuel oil is classified in five standard numbers:

> Oil Nos. 1 and 2 are the lighter and more expensive.
>
> Oil No. 3 is heavier than oil Nos. 1 and 2.
>
> Oil No. 4 is not commonly used.
>
> Oil Nos. 5 and 6 are the heavier and less expensive.
>
> **Oil Nos. 5 and 6 are used in power plants to produce steam.**

## E-13   THE AMOUNT OF BTU PRODUCED BY FUEL OIL:

| Fuel oil Standard No. | Btu produced per gallon |
|---|---|
| 1 | 136,000 |
| 2 | 138,000 |
| 3 | 141,000 |
| 4 | not commonly used |
| 5 | 148,000 |
| 6 | 152,000 |

**with efficiency of 75 percent.**

## E-14   NATURAL GAS

Natural gas is a by-product or an end product of the evolution of petroleum.

Although it is associated with the production of petroleum, it is found in sand, limestone, and sandstone far away from petroleum fields.

Natural gas is a mixture of gases issuing from the ground or from driven wells.

Its composition consists of 80 to 95 percent methane and 5 to 20 percent carbon dioxide, hydrogen, carbon monoxide, nitrogen, helium, and hydrogen sulphide.

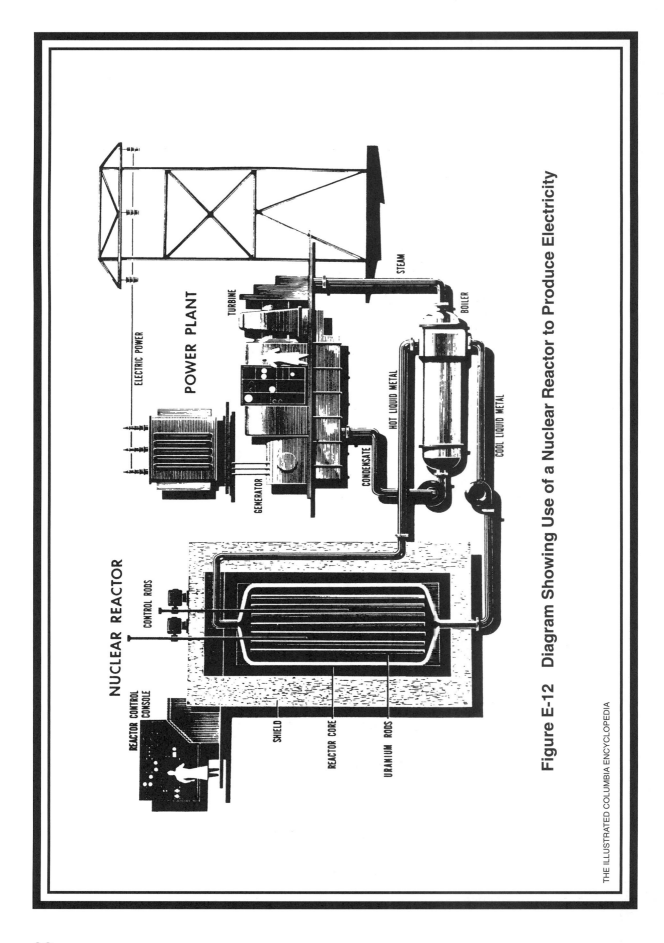

**Figure E-12   Diagram Showing Use of a Nuclear Reactor to Produce Electricity**

Early in the 19th century harnessing the natural gas began in New York State, and by the end of the 19th century natural gas was in use in many industrial cities in the United States and other countries.

**One cubic foot of natural gas produces 1052 Btu with 80 percent efficiency.**

**E-15    NUCLEAR REACTOR TURBINE**

A nuclear reactor is a system used to produce nuclear energy.

In a power plant the heat produced by the reactor is used to produce steam for the turbine.

In a reactor the fission of atomic nuclei produces a self-sustaining nuclear chain reaction in which the neutrons produced are able to split other nuclei.

A chain reaction can be produced in a reactor by using *uranium* or *plutonium* in which the concentration of fissionable isotopes has been artificially increased.

Fissionable isotopes capture adequate high-velocity neutrons to make possible a self-sustaining chain reaction.

The heat energy released by fission in the reactor heats the liquid or gas coolant. This coolant circulates in and out of the reactor core and is radioactive.

The coolant circulates through a heat exchanger outside the reactor core, where the heat energy is transferred to another medium which is nonradioactive and carries the heat energy to be used for producing steam for the turbine (Fig. E-12).

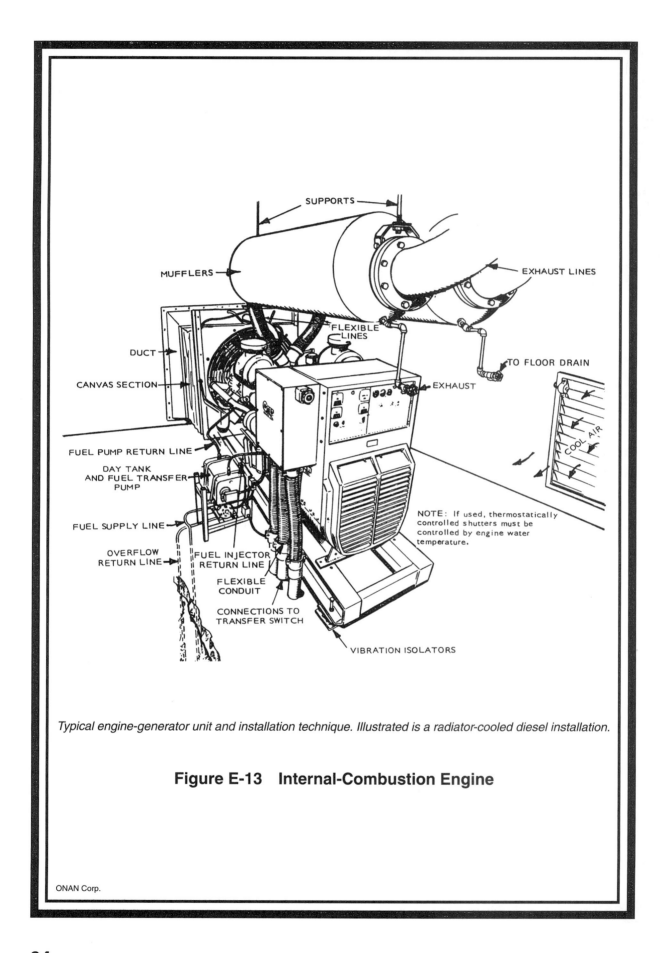

SUPPORTS

MUFFLERS

EXHAUST LINES

FLEXIBLE LINES

DUCT

TO FLOOR DRAIN

CANVAS SECTION

EXHAUST

COOL AIR

FUEL PUMP RETURN LINE

DAY TANK
AND FUEL TRANSFER
PUMP

NOTE: If used, thermostatically
controlled shutters must be
controlled by engine water
temperature.

FUEL SUPPLY LINE

OVERFLOW
RETURN LINE

FUEL INJECTOR
RETURN LINE

FLEXIBLE
CONDUIT

CONNECTIONS TO
TRANSFER SWITCH

VIBRATION ISOLATORS

*Typical engine-generator unit and installation technique. Illustrated is a radiator-cooled diesel installation.*

## Figure E-13   Internal-Combustion Engine

ONAN Corp.

# PRODUCTION OF ELECTRICAL ENERGY
# WITHIN A PROJECT

### E-16 INTERNAL-COMBUSTION ENGINE (ICE)

Internal-combustion engines are used in buildings for two purposes:

1. **Where the supply of electricity is not available.**
2. **For emergency standby power.** Upon the interruption of normal electric power, the ICE starts operating to supply electric power to equipment essential for the safety of the occupants of a building.

The operation of ICE is identical to the engine of an automobile, which rotates a shaft connected to the wheels of the car.

In ICE the shaft rotates the generator for producing electricity (Fig. E-13).

The internal-combustion engine was invented by **Rudolf Diesel** in 1892 using coal dust as fuel. The diesel engine is named after him.

Today a diesel engine burns what was known to be low-cost fuel oil. It does not require a long warming-up period or a large water supply, and it is the most efficient engine for converting heat energy to mechanical energy.

The difference between the ICE and the gasoline engine of a car is that the ignition in an ICE is caused by compression of air in its cylinders. The ignition in a gasoline engine is caused by sparks from spark plugs.

### E-17 WIND TURBINE

A wind turbine is a machine that converts the power of wind to electrical power.

In the 12th century windmills became known in Europe, and today they are landmarks in Holland (Fig. E-14).

Windmills were (and still are) used to grind grains, pump water, etc.

In the 1930s wind turbines were used extensively to produce electricity in remote homes and farms in the United States.

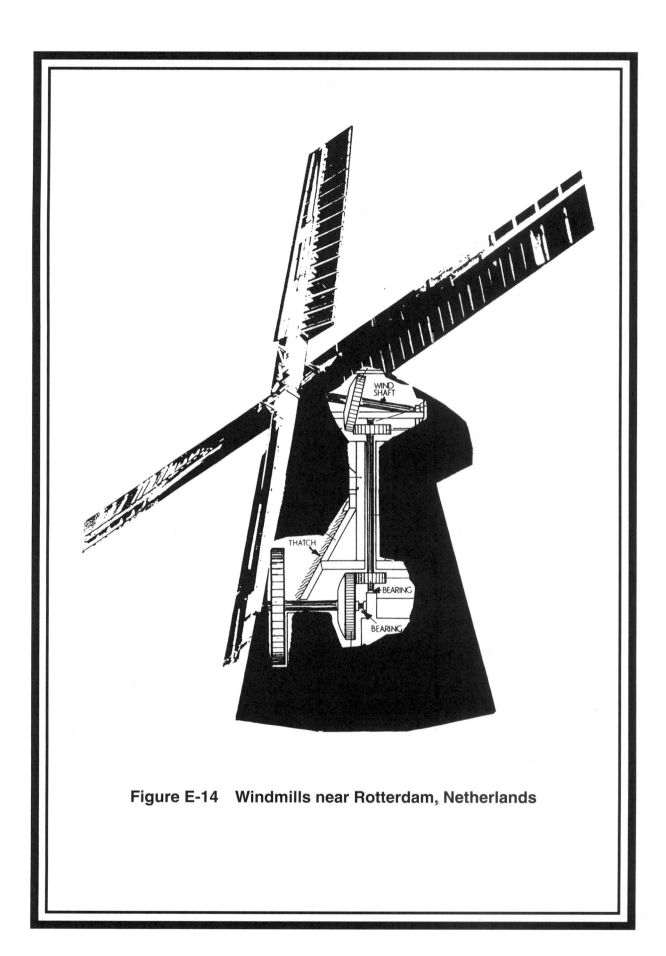

**Figure E-14    Windmills near Rotterdam, Netherlands**

In the 1950s, the spread of the *rural electrification program* in the United States replaced the wind turbines, because the cost of using a public electric system was cheaper, more plentiful, and always available.

Once the initial capital investment is made to purchase and install the wind turbine, the operation cost is essentially free (Fig. E-15).

The cost of a wind turbine and its installation is very high in proportion to the electricity it produces. That is why it is not used extensively.

A wind turbine requires at least 10-mile-per-hour winds to operate properly.

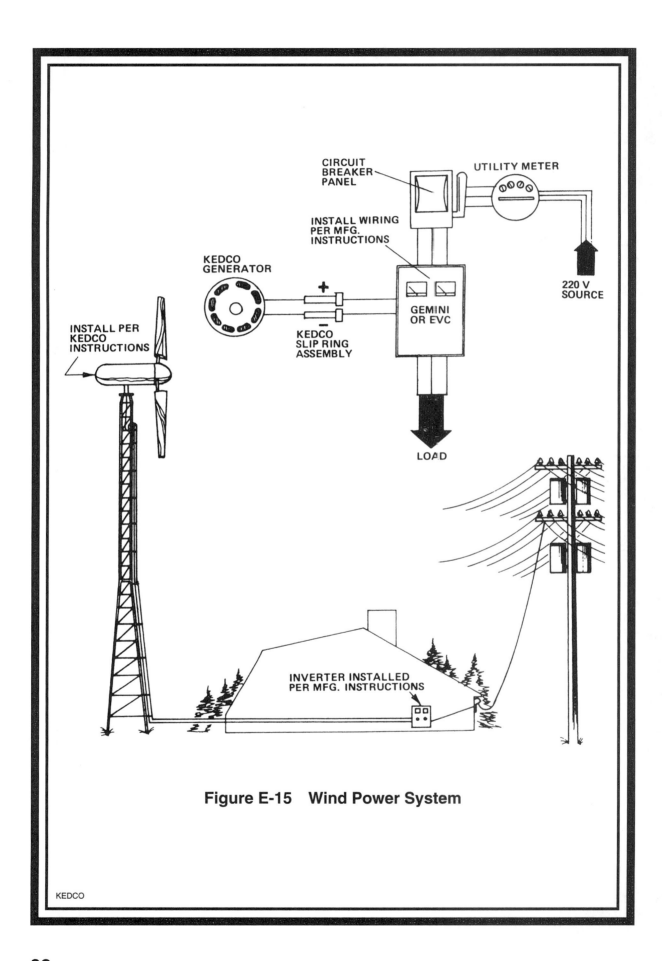

**Figure E-15   Wind Power System**

KEDCO

98

## E-18    WIND AND WIND TURBINE

Wind is the movement of air from high-pressure areas to low-pressure areas. The greater the difference in pressure, the stronger the wind.

The direction of the wind is not always the same near the surface of the earth as it is some miles above the earth.

Wind speed is always higher above the ground; wind power is proportional to the wind speed cubed.

To calculate the total wattage obtained from a wind turbine, the following formula is used:

$$W = 0.0024\ D^2\ V^3\ E$$

where   $W$ = watts

$D$ = diameter of the propeller in feet

$V$ = wind velocity in miles per hour (mph)

$E$ = efficiency due to friction and other limitations ($E$ = 75 to 85%)

### Example

The wind turbine has a propeller 7 feet in diameter. If the wind velocity is 20 mph and $E$ = 80% how many watts are produced?

### Solution

$W = 0.0024\ D^2\ V^3\ E$

$W = 0.0024\ (7)^2\ (20)^3\ (0.8) = 768$ watts

**Figure E-16  Photovoltaic Cells**

# E-19 SOLAR PHOTOVOLTAIC CELLS

Harnessing the sun's power to produce electricity was developed for the space program in 1954.

**Solar cells** convert sunlight directly into electrical power. They are usually made of a small wafer of silicon into which is dispensed a limited amount of arsenic. The wafer is treated with boron gas, producing on its surface a modified gas zone.

The contact between the wafer and modified gas zone produces voltage which flows by the action of the sunlight.

**All silicon cells, regardless of size, produce 0.45 volts per cell.**

The current produced by silicon cells depends directly on the direction and the amount of sunlight it receives. The maximum current of silicon cells is normally 1.2 amperes or more (Fig. E-16).

The cells are commonly connected to batteries and load. When the sun is shining, cells will charge the batteries and power the load, and when the sun is not shining on the cells, the batteries will power the load (Fig. E-17).

The costs of installation of solar cells are extremely high in proportion to the cost of the electricity they produce.

To install the solar cells, the cost is approximately $45 per watt or $45,000 per kilowatt it produces. The average cost of electricity is $0.12 per kWh. Therefore, it will take several hundred years to pay back the initial cost. For this reason its use today is not economically feasible.

**Figure E-17    Concentration of Sunlight on a Photovoltaic Cell**

# GENERATING ELECTRICITY

**E-20 GENERATOR**

A generator is simply a mechanical arrangement for producing a current at a specific voltage (Fig. E-18). To accomplish this task it is designed to:

1.  Rotate *armature* between the poles of a permanent magnet in smaller generators, or rotate the magnetic field called *field winding* past the armature in large generators (Fig. E-19).
2.  Produce current and voltage as follows:
    *a.* The armature iron is subject to varying magnetic flux, and by a whirling motion, *current* will be induced in it. (To reduce an eddy-current loss the armature iron is built up of thin laminations.)
    *b.* The armature-generated *voltage* results from the mechanical motion of either the armature or the magnetic field.
3.  Lead the current and voltage to an external circuit.

There are two types of electric energy produced by generators:

1.  *Direct current* (DC or dc)
2.  *Alternating current* (AC or ac)

Direct current is produced in a generator when a *split ring commutator* is used to lead the current and voltage to an external circuit (Fig. E-20).

Alternating current is produced in a generator when a *split ring* is used to lead the current and voltage to an external circuit (Fig. E-21).

**E-21 DIRECT CURRENT (DC or dc)**

Direct current was advocated by Thomas A. Edison in 1879, and the first central electric light power plant in the world—Pearl Street Plant—was completed by Edison in New York City in 1882. To this date many structures in lower Manhattan are served with direct current, because the electrical wiring, devices, and equipment have been designed for dc, and converting them to ac is costly.

Direct current is produced by **dc generators,** with a constant current flow rate induced by a constant voltage.

*Sectional view of a steam-turbine-driven synchronous generator. The ac armature winding is on the stator. The dc field winding is on the cylindrical rotor.*

**Figure E-18   Synchronous Generator**

*Dc generator or motor armature in process of being wound. One side of each coil is placed in the bottom of a slot; the other side is placed in the top of a slot.*

**Figure E-19   Motor Armature**

The current produced is at a constant time rate and always in the same direction around a circuit (Fig. E-22).

In the flow of electricity in a circuit, one wire is always positive and the other is always negative. There is no frequency in dc, because the voltage never changes its polarity; therefore, transformers cannot be used for dc.

Direct-current resistance in a circuit is called **resistance—R**. The Ohm's law for dc is expressed as follows:

$$I = \frac{V}{R} \quad \text{and} \quad R = \frac{V}{I}$$

where $I$ = current (in amp)    $V$ = volt (in volt)    $R$ = resistance (in ohm)

All batteries are charged *only* with direct current.

## E-22    WHY DC IS NOT COMMONLY USED

When the first dc power plant in the world became operational in New York City, at the same time "**anti-dc**" was invented, the "*automobile*" allowed the rapid expansion of towns and cities.

When dc is generated, it has to be utilized in the same instant.

The high current at 240 V produced at the power plant has to be sent through the service conductors to the consumers (with no transformers).

**High current requires:**

1.    **Conductors with a large diameter, which are costly, and**
2.    **Power loss in the conductors can be tolerated for a short distance.***

For this reason direct current is not commonly used.

* When current moves through the conductors, it creates heat. Therefore, electric energy is converted to heat energy and dissipates in the surrounding air (power loss in the line). For example, if 240,000 kW of dc is sent 20 miles away from a power plant through service conductors, probably very little or no power is received at the end of the service conductors.

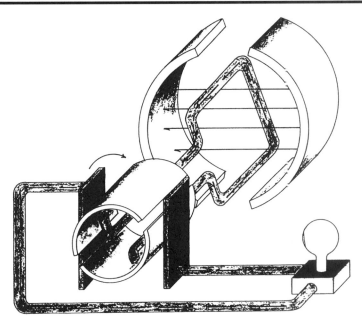

*In the second half of the turn, right, the current in the armature reverses direction and flows out of commutator segment. But current in the outside circuit remains the same.*

**Figure E-20    Direct Current Generator**

*When the armature rotates into the magnetic field again, right, the current reverses direction. It now flows out of the armature through slip ring and back into the armature at slip ring.*

**Figure E-21    Alternating Current Generator**

### E-23    USE OF DIRECT CURRENT IS LIMITED

Direct current has a constant current flow rate induced by a constant voltage. When it is utilized in a dc motor it produces exact revolutions per minutes (RPM). **Direct current is used for many applications which require "proper" RPM to perform properly, such as:**

1.    **Vertical transportation**

2.    **Rapid-transit propulsion systems**

3.    **Vehicles**

4.    **Electrochemical processing**

5.    **Printing-press drives**

6.    **Electroplating**

7.    **Emergency lighting, controls and communication, etc.**

8.    **Motors and compressors in HVAC**

Since the electricity available is in the form of alternating current (ac), we use dc generators of proper size driven by ac motors to obtain dc power in the buildings.

Small quantities of dc are produced by batteries or rectifiers for emergency lighting, communication, signals and control equipment, etc.


### E-24    ALTERNATING CURRENT (AC or ac)

Nikola Tesla, after receiving patents on his inventions of the ac induction motor and generator of high-frequency current, sold his patents to George Westinghouse.

Westinghouse, in partnership with Tesla, constructed the hydroelectric power plant at Niagara Falls. In 1890, electricity generated was used in Buffalo, New York, 30 miles away from the power plant. This was the beginning of the use of alternating current in the United States and abroad.

Alternating current is produced by an **ac generator** commonly known as an *alternator.*

**The alternator works on the same basic principle as the dc generator.**

In ac circuits, the electrons are causing the current to move in one direction, then in the reverse direction; therefore, the current changes direction many times per second.

That is why it is called **alternating current.** The changes from (+) half cycle to (−) half cycle and back to (+) half cycle constitute a full cycle.

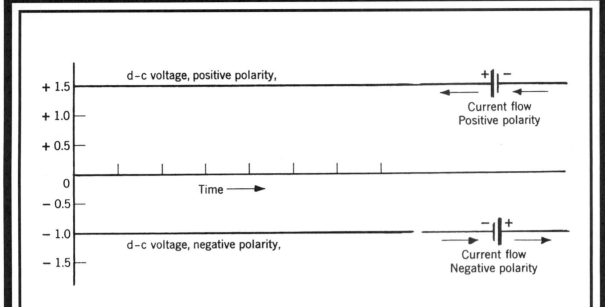

**Figure E-22   Graphical Representation of D-C Voltage with Positive and Negative Polarity**

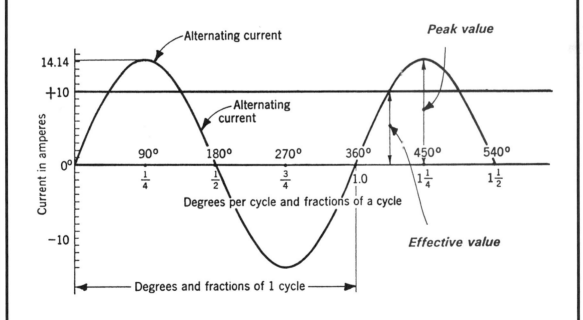

**Figure E-23   Effective Value of Alternating Voltage**

Alternating current works in both the positive and negative half cycle, and the reversing process does not impede the energy transfer.

As Fig. E-23 shows, the valve of voltage follows a sine wave curve with respect to time; the resulting current follows the same type of curve.

**E-25   EFFECTIVE VALUE OF CURRENT AND VOLTAGE IN AC**

As shown in Fig. E-23, there are two voltage values:

1. **Effective value**
2. **Peak value**

   Effective value = peak value × 0.707

The effective value is the quantity of $VI$ that would produce the heating effect in resistance.

The effective value of $VI$ is the usable power and energy.

**In all electrical calculations, only effective values of $VI$ are used.**

   $V$ = volt      $I$ = current

**E-26   DISTRIBUTION OF AC**

When ac is generated, it has to be utilized in the same instant.

Since the ac moves through conductors with a speed of 180,000 miles per second, there is no problem in accomplishing this goal.

Voltage and current generated in the power plant are transferred to a ***step-up* transformer.**

***VI*** entering the transformer is equal to ***VI*** leaving the transformer, with nominal power loss within the transformer.

**A step-up transformer** owers the current and increases the voltage approximately 138,000 V.

The amount of current leaving the transformer is very low; therefore, it requires thin conductors.

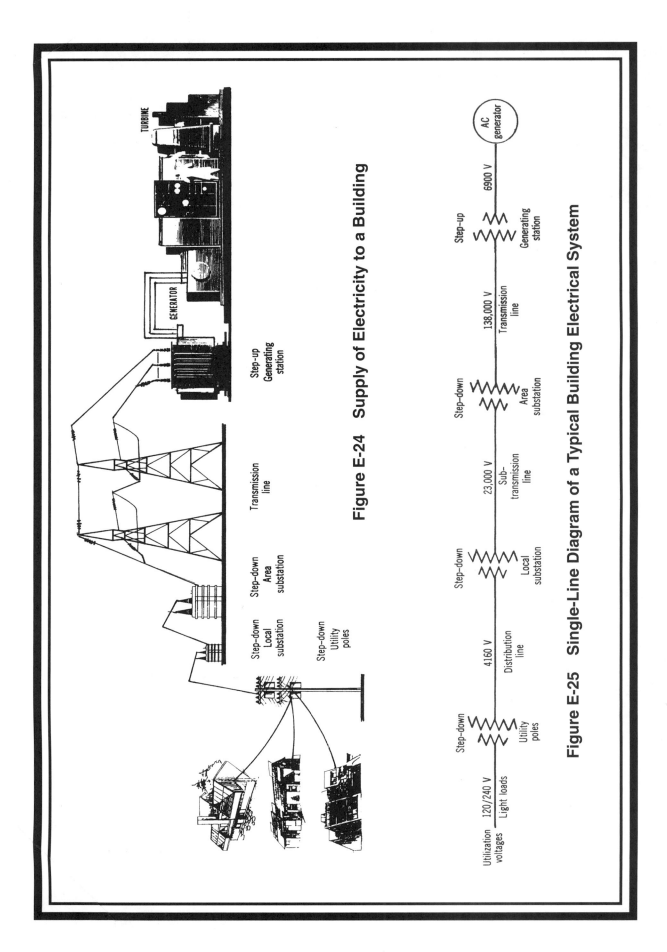

**Figure E-24  Supply of Electricity to a Building**

**Figure E-25  Single-Line Diagram of a Typical Building Electrical System**

High voltage and low current is transported through thin *high-voltage* **power lines** to the area where **VI** is to be utilized.

The high-voltage power lines enter into the area substation *step-down* **transformer,** which lowers the voltage and raises the current.

An area distribution service line with **4160 V carries the VI** for use by consumers.

There is a **step-down transformer** for each structure or group of structures with a capacity required to produce 120/240 V, 120/208 V, 277/480 V, and 2400/4160 V (Figs. E-24 and E-25).

Therefore, an ac power plant can be located in the area where it is more economical to produce electricity and transport the **VI** hundreds of miles away to the consumers.

## E-27    FREQUENCY (F) "for AC only"

Frequency is created only in alternating current because of the behavior of voltage and current.

Frequency is the **number of cycles** of alternating voltage or current per second. It is also called **hertz.**

**Cycle = hertz (Hz)**

Frequency of the voltage or current is directly related to the design of the generator, its number of poles, and the speed at which it is driven.

**In the United States alternating current is produced in 60 Hz.**

In Europe and some other countries ac is produced in 50 Hz.

In Eastern Europe and some other countries ac is produced in 25 Hz. This low cycle creates a flicker in incandescent lamps and other devices.

**Equipment made for one frequency does not work properly with other frequencies, because:**

1.    **They will overheat.**
2.    **They may burn out.**
3.    **They will not perform as they should.**
4.    **They will have a shorter life.**

# E-28 DC AND AC AT A GLANCE

## DIRECT CURRENT—DC

Is produced by a DC generator with an attachment called **commutator,** which rectifies the AC to DC (Fig. E-20).

Electric current flows at a constant time rate and in the same direction. Therefore, it has no frequency (Fig. E-22).

Electricity produced in a power plant is sent directly with high current through thick conductors to consumers.

Transformers cannot be used in DC.

Power produced can serve consumers at a short distance.

Is used for special application, such as vertical transportation, motors, and communication. (Please see E-23.)

## ALTERNATING CURRENT—AC

Is produced by an AC generator called an **alternator** (Fig. E-21).

Electric current flows in varying time and direction along the time axis. It flows with (+) and (−) loops, creating a cycle; the number of cycles per second is called **frequency** (Fig. E-23).

**Cycle = hertz Hz**

**In the United States, AC is produced at 60 Hz.**

Electricity produced in a power plant is sent to a nearby "step-up" transformer, and electricity with low current and high voltage is sent through thin conductors (high-voltage power line) to consumers.

Transformers are used in AC.

Power produced can serve consumers hundreds of miles away.

The bulk of electricity used today is in the form of AC.

| DIRECT CURRENT—DC | ALTERNATING CURRENT—AC |
|---|---|
| When used in motors, it produces a precise revolution per minute (RPM). | When used in motors, it does not produce a precise revolution per minute (RPM). |
| DC can be stored in batteries. **All batteries are charged by DC.** | AC cannot be stored in batteries. |
| Resistance is called *resistance* (R) | Resistance is called *impedance* (Z). |

Direct Current—DC:

$$I = \frac{V}{R} \text{ for all calculations}$$

$$W = VI \text{ for all calculations}$$

Alternating Current—AC:

$$I = \frac{V}{Z} \text{ for all calculations except power equipment}$$

$$W = VI \text{ for appliances with resistive elements only}$$

$$W = VI \times PF \text{ for calculations of power equipment}$$

# ELECTRIC SYSTEMS COMPONENTS AND CALCULATIONS

**E-29**    **OHM'S LAW**

Ohm's law states that when **a current (*I*)** flows through a circuit with a given **resistance (*R*)**, it is directly proportional to the **voltage (*V*).**

$$I = \frac{V}{R}$$

For resistive loads such as heaters, toasters, irons, and incandescent lights (*not* including power equipment):

$$\text{dc resistance } (R) = \text{ac resistance } (Z)$$

Therefore,

$$R = \frac{V}{I} \text{ for dc} \qquad \text{and} \qquad Z = \frac{V}{I} \text{ for ac}$$

*Example*

A dc heater draws 10 amps at 240 volts. What is the resistance of its heating element?

*Solution*

$$R = \frac{V}{I} \qquad R = \frac{240 \text{ volts}}{10 \text{ amps}} = 24 \text{ ohms}$$

*Example*

An ac heater draws 10 amps at 120 volts. What is the resistance of its heating element?

*Solution*

$$Z = \frac{V}{I} \qquad Z = \frac{120 \text{ volts}}{10 \text{ amps}} = 12 \text{ ohms}$$

where    *I* = ampere    *V* = volt    *R* = resistance    *z* = impedence

**114**

## E-30  CURRENT (UNIT: AMPERE) "for AC or DC"

The flow of ac or dc in a conductor is called **current** or **amperage,** and is designated by symbols (A, a, or amp) and represented in an equation by the letter *I.*

In a generator armature  iron is subject to carrying magnetic flux, and by a whirling motion **current** will be induced in it.

**The unit of measurement of the current is** *ampere,* **named after André M. Ampère.** (Please see G-10.)

One ampere is used when $6.251 \times 10^{18}$ free electrons pass a given cross section in 1 second. (Please see E-3 and E-4.)

### *Example*

An electric heater is rated 900 W–120 V

a.    What is the current through it?

b.    What is the number of free electrons passing through the cross section of its resistance in 1 second?

c.    What is the cost of operating this heater for 2 hours if the electric cost is 14 cents per kWh?

### *Solution*

a.    $W = \dfrac{V}{I}$      $I = \dfrac{W}{V}$      $I = \dfrac{900 \text{ W}}{120 \text{ V}} = 7.5$ amps

b.    7.5 amps $\times 6.251 (10)^{18} = 46.89 (10)^{18}$ free electrons

c.    900 W $\times$ 2 h = 1800 Wh = 1.8 kWh

1.8 kWh $\times$ 14 cents/kWh = 25.2 cents

where   W = watt   h = hour   k = kilo = 1000 W

**E-31  VOLTAGE (UNIT: VOLT) "for AC or DC"**

It is also referred to as *electrical potential, potential difference,* and *electromotive force* **(emf)** and is designated by symbol **(V).**

In a generator the armature-generated *voltage* results from a mechanical motion of either the armature or the magnetic field.

**The unit of measurement of voltage is *volt,* named after Allesandro G. Volta.** (Please see G-11.)

Between positive and negative terminals, there is a force created by the charged free electrons. This force is called *electromotive force* **(emf). The emf** causes the current to flow through the conductors. Therefore, the higher the voltage **(emf),** the higher the current that will flow through the conductor.

**E-32  RESISTANCE (for DC or AC)**

*Resistance* **is an electrical term for friction.**

The flow of current in an electrical circuit is resisted, impeded by resistance.

Any device whose electric characteristic is resistance is called a *resistor.*

Resistance opposes the current, and the energy dissipated because of this opposition appears as *heat.*

When an electric charge gives up energy by passing through a resistance, a voltage drop appears in the direction of the current.

The resistance of materials used in electrical circuits depends directly on the numbers of free electrons they contain.

The higher the number of free electrons, the lower the resistance, such as in copper and aluminum conductors.

The lower the number of free electrons, the higher the resistance in *insulators* (rubber, glass, and synthetic materials, etc.).

**The unit of measurement for resistance is *ohm,* named after Georg S. Ohm.** (Please see G-9.)

**One ohm is defined as the resistance which allows 1 amp of current to flow through it with a pressure of 1 volt.**

In dc circuits resistance is called *resistance (R)*.

In ac circuits resistance is called *impedance (Z)*.

Impedance is a compound of resistance plus *reactance*.

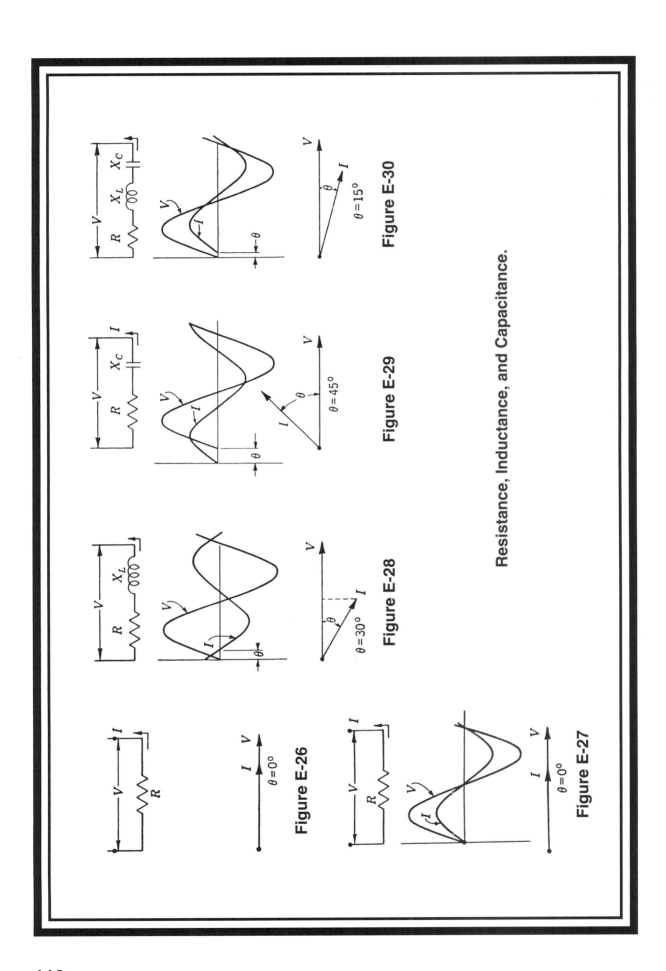

Figure E-30

Figure E-29

Figure E-28

Figure E-27

Figure E-26

Resistance, Inductance, and Capacitance.

118

## E-33    DIRECT CURRENT RESISTANCE (R)

In direct current circuits, the current and voltage are always in a phase (Fig. E-26).

Therefore,

$$R = \frac{V}{I} \quad \text{and} \quad W = VI, \quad W = RI^2$$

## E-34    ALTERNATING CURRENT IMPEDANCE (Z)

Impedance is a component of resistance plus *reactance.*

In **AC** circuits, there are four conditions for the use of impedance **(Z):**

1.  If an **AC** circuit contains only resistive appliances such as heaters, irons, toasters, and incandescent lamps, the current and voltage vectors are in phase (Fig. E-27); then

    **DC resistance (R) = AC impedance (Z)**

    $$Z = \frac{V}{I} \quad \text{and} \quad W = VI, \quad W = ZI^2$$

2.  When an **AC** circuit contains resistance **(R)** and *inductive reactance* **($X_L$)**, the **($X_L$)** causes the current to lag the voltage by an angle, the cosine of which is called *power factor* **(PF)** (Fig. E-28). In this case,

    $$Z = \sqrt{R^2 + X_L^2} \quad \text{and} \quad W = VI \times \text{PF}$$

3.  When an **AC** circuit contains resistance **(R)** and *capacitative reactance* **($X_c$)**, the **($X_c$)** causes the current to lead the voltage by a **PF** (Fig. E-29). Then

    $$Z = \sqrt{R^2 + X_c^2} \quad \text{and} \quad W = VI \times \text{PF}$$

4.  When an **AC** circuit contains resistance **(R)**, inductive reactance **($X_L$)**, and capacitative reactance **($X_c$)** (Fig. E-30), the following conditions will occur within the circuit:

    a.  If **($X_L$)** is greater than **($X_c$)**, **the current lags the voltage.**

    b.  If **($X_c$)** is greater than **($X_L$)**, **the current will lead the voltage.**

    In this case

    $$Z = \sqrt{R^2 + (X_L - X_c)^2} \quad \text{and} \quad W = VI \times \text{PF}$$

## E-35    POWER (UNIT: WATT)

**Power is the time rate for doing work or the rate at which energy is used.**

**Work = energy = power × time**

In the English system of measurement units of power are:

1.    **Btu**

2.    **Horsepower**

3.    **Watt**

4.    **Kilowatt**

## E-36    BRITISH THERMAL UNIT (Btu) AND HEAT POWER

British thermal unit is:

**the amount of heat power needed to raise the temperature of 1 pound of water 1°F**

*Heat power* is obtained by using fuel or electricity. The amount of Btu produced is as follows:

| | | |
|---|---|---|
| **One pound of coal** | **14,600 Btu** | **with 70% E.** |
| **One gallon of oil No. 6** | **148,000 Btu** | **with 75% E.** |
| **One cubic foot of natural gas** | **1,052 Btu** | **with 80% E.** |
| **One watt of electricity** | **3.41 Btu** | **with 100% E.** |

E = efficiency

## E-37    HORSEPOWER (HP)

Horsepower was devised for use in mechanics by James Watt.

Horsepower is based on the concept that a horse can pull 550 foot-pounds per second.

*Foot-pound* is the work done when 1 pound is moved a distance of 1 foot.

In electricity all equipments are rated in horsepower:

**HP = 746 W**

**E-38    WATT (W)**

**The unit of power is *watt,* named after James Watt.** (Please see G-12.)

**Watt is when 1 ampere is flowing under the force of 1 volt and is expressed in watts (W).**

The resistive loads of all electrical devices (heater, toaster, iron, incandescent lights, etc.) are the same in direct current and alternating current; therefore, the calculation for power (W) is as follows:

$$\textbf{Resistance } (R) = \textbf{impedance } (Z)$$
$$W = VI \textbf{ for DC and AC}$$

For power equipment the calculation of power (*W*) for DC differs from AC as follows:

$$W = VI \qquad \textbf{for DC}$$
$$W = VI \times \textbf{PF} \qquad \textbf{for AC}$$

where    $W$ = watt    $V$ = volt    $I$ = ampere    PF = power factor

**E-39    POWER FACTOR (PF) (for AC only)**

Impedance (*Z*) causes a phase difference between current and voltage.

This phase difference is represented by an angle, the cosine of which is called ***power factor (PF).***

The power factor angles of any equipment vary from 0° to 90° lagging or leading, causing a reduction in percentage of power used.

In all types of equipment, a percentage of PF is given for the calculation to determine the power requirements.

The equipment with high PF is more efficient and more costly.

The equipment with low PF is less efficient and less costly.

$$\text{Power efficiency} = \frac{\text{power output}}{\text{power input}}$$

$$1 \text{ HP} = 746 \text{ W}$$

$$\text{Power output} = \text{number of HP} \times 746 \text{ W/HP}$$

$$\text{Power input} = \frac{\text{power output}}{\text{power efficiency}} \quad \text{or} = V \times I \times PF$$

$$\text{Power factor (PF)} = \frac{\text{power input}}{VI}$$

$$\text{power input} = V \times I \times PF$$

### Example

A 4-HP ac motor is rated for 240 V and 20 amp. Assuming efficiency of the motor is 85 percent.

a.    What is power output?

b.    What is power input?

c.    What is power factor PF?

### Solution

a.    Power output = 4 HP $\times$ 746 W/HP = 2984 W

b.    Power input = $\dfrac{2984 \text{ W}}{0.85}$ = 3510.59 W

c.    Power factor = $\dfrac{3510.59 \text{ W}}{240 \text{ V} \times 20 \text{ amp}}$ = 0.74%

**122**

## E-41   MEASUREMENT OF ELECTRICAL ENERGY

The amount of electrical energy used is proportional to the power (W) used in the building and the duration of time it consumed.

**Electrical energy used = power (W) × time (h)**

Unit of energy is watt-hour (Wh) or kilowatt-hour (kWh).

Kilo is taken from the metric system and means "1000."

Since the number of watts used is large, it is divided by 1000 to produce (kW):

$$kW = \frac{W}{1000}$$

### Example

Find the total cost of electrical energy used in one month (31 days) if the cost of electricity is $0.12 per kWh for the following uses:

| | |
|---|---|
| Six 100-W incandescent lights | used 8 h/day |
| One 1200-W toaster | used 0.25 h/day |
| One 1000-W iron | used 1.0 h/day |
| Two 1500-W heaters | used 6.0 h/day |

### Solution

| | |
|---|---|
| Inc. lamps | $6 \times 100$ W $\times$ 8 h/d = 4800 W h/d |
| Toaster | $1 \times 1200$ W $\times$ 0.25 h/d = 300 W h/d |
| Iron | $1 \times 1000$ W $\times$ 1 h/d = 1000 W h/d |
| Heaters | $2 \times 1500$ W $\times$ 6 h/d = 18,000 W h/d |

4800 W h/d + 300 W h/d + 1000 W h/d + 18,000 W h/d = 24,100 W h/d

24,000 W h/d ÷ 1000 W/KW = 24.1 kWh/d

24.1 kWh/d $\times$ 31 d/m = 747.1 kWh/m

747.1 kWh/m $\times$ $0.12/kWh = $89.65 per month

where  h = hour   d = day   m = month

## E-42   BATTERIES

Peter von Musschenbroek in 1774 invented the *Leyden jar,* which stored static electricity.

William Watson discharged a Leyden jar through a circuit, and the comprehension of the current and circuit started a new field of experimentation.

The invention of the battery was the base for the discovery of G. S. Ohm's law.

When two different metals, such as copper and zinc, are placed in a container filled with salt, water, or diluted sulfuric acid, a current is caused to flow between the electrodes.

There are four types of batteries used today:

1.   **Dry-cell batteries**

2.   **Storage batteries (wet-cell)**

3.   **Edison cell battery (wet-cell)**

4.   **Solar battery**

## E-43   DRY-CELL BATTERIES

Dry-cell batteries are a group of dry cells connected to act as a source of direct current at a given voltage. They are used for battery-operated devices.

They are designated by letters according to their function:

> **The "A" battery provides 1.5 to 12 volts.**
>
> **The "B" battery provides 6 to 12 volts.**
>
> **The "C" battery provides 1.5 volts.**

## E-44   STORAGE BATTERIES (WET-CELL)

They consist of several cells connected in series. Each cell contains a number of alternately positive and negative plates separated by the **liquid electrolyte** of a sulfuric acid solution.

The (+) plates of the cell are connected to form the (+) electrode.

The (−) plates of the cell are connected to form the (−) electrode.

When it is charged, the cell is made to operate in reverse of its discharging operation, causing the reverse of the chemical reaction, so that **electrical energy** is converted into **stored chemical energy.**

In the United States and some other countries the *lead storage* battery is commonly used.

In Europe and some other countries a *nickel cadmium* battery is widely used.

Storage batteries are used for emergency lighting, power for standby electric service, telephone equipment power, railway signaling, transportation applications, etc.

In large structures a bank of batteries, called *uninterruptible power system* **(UPS),** is used to provide adequate power when power interruption could cause disastrous consequences (Fig. E-31).

## E-45    EDISON CELL BATTERY (WET-CELL)

Edison cell battery has a nickel oxide positive plate and an iron negative plate suspended in a solution of potassium and lithium hydroxides. It has a longer effective life and the capacity to withstand abuse. It is used mostly in railroads.

## E-46    SOLAR BATTERY

Please see E-12.

**Figure E-31    Storage Batteries (Wet-Cell)**

# CONTENTS

# Part 3

## ELECTRICAL SYSTEMS

# S

## Electrical Systems

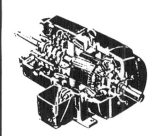

## Electrical System Voltage

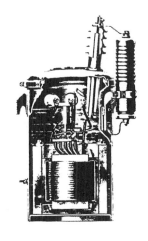

## Transformer System

# Electrical Motors

# ELECTRICAL SYSTEMS

## S-1 ELECTRICAL SYSTEMS AC

Electrical systems for a project include:

1. **Electric services**
2. **Transformers**
3. **Load center, switches, panelboards, and overcurrent protection**
4. **Feeders and branch circuits**
5. **Electrical equipment, lighting, and electrical heating**
6. **Electric motors and controllers**
7. **Miscellaneous electrical systems**
8. **Signaling circuits**

## S-2 CODES

The safety of the electrical systems installed and maintained in the building(s) is of **prime importance.**

All electrical codes define the *minimum* fundamental safety requirements that *must* be followed in:

1. **Design and preparation of construction documents**
2. **Selection**
3. **Construction**
4. **Installation**

of all electrical systems.

Before designing the electrical system, you *must* determine what code is applicable to the area in which the structure will be built. Many large cities have their own electrical codes.

Normally they are similar to the **National Electric Code (NEC),** with certain changes or additions to fulfill their special requirements. Other cities, towns, or villages may use the state or the large cities' electrical code or **NEC.**

If the structure(s) is to be constructed in a region or a country which has no code requirements, you should use **NEC** or other codes used by large cities in the United States.

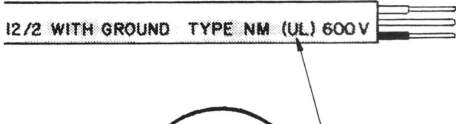

**Figure S-1    Typical Underwriters Laboratories Seal**

| System Voltage (Transformers) | | Utilization Voltage (Motors) | |
|---|---|---|---|
| Nominal | With 4% Drop | Current Standard | Obsolete Standard |
| 120 | 115.2 | 115 | 110 |
| 208 | 199.7 | 200 | 208 |
| 240 | 230.4 | 230 | 220 |
| 480 | 460.8 | 460 | 440 |
| 600 | 576.0 | 575 | 550 |

**Figure S-2    System and Utilization Voltages**

These codes have been written to protect and safeguard the occupants and users of building(s), as well as to protect the architects who design them.

A safe design awards us with

**a peace of mind and lifetime protection against lawsuits**

### S-3   NATIONAL ELECTRICAL CODE

The National Electrical Code **(NEC)** of **the National Fire Protection Association (N.F.P.A.)** defines the minimum safety requirements that must be followed in the selection, construction, and installation of all electrical systems. A copy of the latest edition of the NEC should be available in your reference library.

Basically there are two rules in the **NEC:**

1.   *Mandatory rules.*   All mandatory rules refer to the word *shall,* which indicates that you must comply with it.
2.   *Advisory rules.*   All advisory rules refer to the word *should* which indicates that it is recommended, but it is not required or mandatory.

*Note:*  Some local electrical codes using **NEC** have amended **NEC** terminology and use *shall* in place of *should.* This change of **"recommended"** to **"mandatory"** governs for that locality only.

### S-4   UNDERWRITERS LABORATORIES

The Underwriters Laboratories **(U.L.)** is an independent organization.

**U.L.** tests, inspects, and approves all electrical systems to make sure they meet minimum specifications as set up by **U.L.**

**U.L.** publishes lists of inspected and approved electrical systems.

Many codes require that the electrical materials used must bear the **U.L.** label (Fig. S-1).

### S-5   LICENSES

Many cities and municipalities have laws which require that the electrician installing the electrical system must be licensed.

This ensures that the electrician is knowledgeable with electrical codes and installation procedures.

In some cities and municipalities there is no law requiring an electrician to be licensed.

In this case the architect **must** make sure that the electrician installing the electrical system has many years of experience and is knowledgeable with the electrical codes.

## S-6    PERMITS

Construction drawings and specifications which include electrical drawings and specifications must be submitted to a municipality for approval in order to receive a construction permit.

Several inspections will be made by an electrical inspector during the installation of an electrical system.

The following procedures are recommended:

1.    Maintain a close coordination with the inspector in order to avoid a slowdown of the construction progress.
2.    Installation of rough wiring system has to be inspected and approved by an inspector before the walls are concealed.
3.    When fixtures, equipment, devices, panel(s) service, and meters are installed they have to be approved.
4.    At every stage or at the end of all installation an approval is normally issued to be posted in the building.

# ELECTRICAL SYSTEM VOLTAGE

## S-7    ELECTRICAL SYSTEM VOLTAGE

The systems of *voltage* commonly used in the United States and Canada are:

| | | | |
|---|---|---|---|
| 1. | 120 V | single-phase | 2-wire |
| 2. | 120/240 V | single-phase | 3-wire |
| 3. | 120/208 V | single-phase | 3-wire |
| 4. | 120/208 V | three-phase | 4-wire |
| 5. | 277/480 V | three-phase | 4-wire |
| 6. | 2400/4160 V | three-phase | 4-wire |

The voltage causes the current to flow through the conductors.

The higher the voltage, the higher the current will flow.

Figure S-2 shows the relationship between voltage and current utilized for motors. As shown, the higher the voltage, the higher the current will be.

## S-8    120-V SINGLE-PHASE 2-WIRE SYSTEM

Used for maximum load of 6 kW (Fig. S-3).

This system is also referred to as 110-V and 115-V.

It is used for a very small building with a total electric load requirement of 50 A.

## S-9    120/240-V SINGLE-PHASE 3-WIRE SYSTEM

This system is also referred to as 115/230-V and 110/220-V (Fig. S-4).

It is used for:

1.    **Residences** up to 5000 sq. ft. with maximum 150-A load

2.    **Stores** up to 2000 sq. ft. with maximum 150-A load

3.    **Schools** up to 5000 sq. ft. with maximum 150-A load

4.    **Churches** up to 2000 sq. ft. with maximum 150-A oad

5.    Other structures with similar requirements

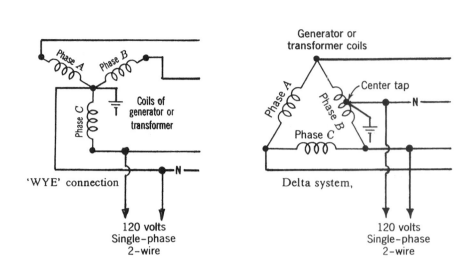

**Figure S-3    120-V Single-Phase 2-Wire**

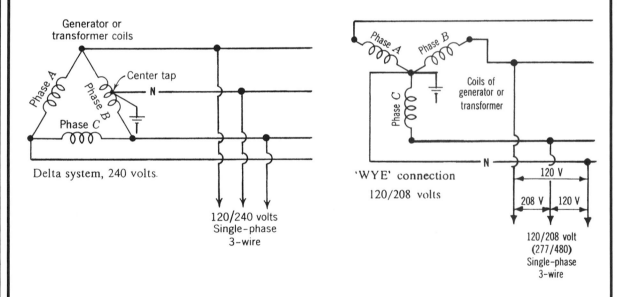

**Figure S-4**                    **Figure S-5**

**120/208-V, 120/240-V, Single-Phase, 3-Wire Service**

**S-10** **120/208-V SINGLE-PHASE 3-WIRE SYSTEM**

This system is similar to 120/240-V single-phase 3-wire (Fig. S-5).

**S-11** **120/208-V THREE-PHASE 4-WIRE SYSTEM**

This system is applicable to buildings with HVAC (Fig. S-6).

1. **Apartment buildings** up to 10,000 sq. ft. with maximum 150 A-load

2. **Hospitals** up to 10,000 sq. ft. with maximum 400-A load

3. **Office buildings** up to 10,000 sq. ft. with maximum 600-A load

4. **Stores** up to 10,000 sq. ft. with maximum 600-A load

5. **Schools** up to 10,000 sq. ft. with maximum 200-A load

**S-12** **277/480-V THREE-PHASE 4-WIRE SYSTEM**

This system is also referred to as 265/460-V and 255/440-V (Fig. S-7).

This system is used in buildings which utilize 277-V fluorescent lighting and 480-V machinery.

Step-down transformers are used to provide 120 V for receptacles and other loads. This system serves multistory buildings and industrial facilities.

**S-13** **2400/4160-V THREE-PHASE 4-WIRE SYSTEM**

This system is used in very large commercial and industrial buildings.

*Note*

1. Transformer voltage standards establish the voltage systems stated supra.

2. Motors have voltage standards which govern the utilization (Fig. S-2).

3. Motors which consume 230 or 240 V should not be used in 120/208, because they will produce excessive heat when they operate and, consequently, will have a shorter life.

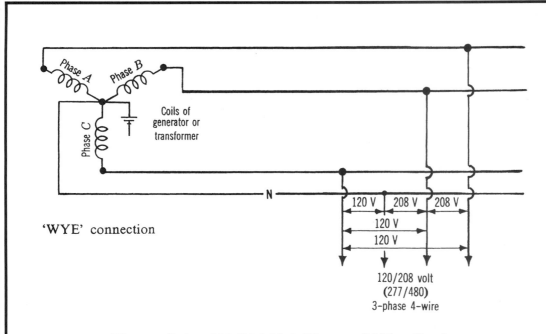

**Figure S-6    120/208-V, 3-Phase, 4-Wire System**

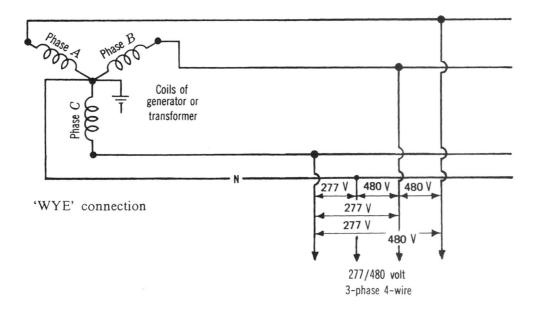

**Figure S-7    277/480-V, 3-Phase, 4-Wire Service System**

# TRANSFORMER SYSTEM

## S-14 TRANSFORMER

It is used only for alternating current **AC.**

A transformer is a device which transforms or nducts **AC** of one voltage to **AC** of another voltage (Fig. S-8).

When a transformer increases the voltage it is called **step-up transformer.**

When a transformer decreases the voltage it is called **step-down transformer.**

A transformer is a **static device,** consisting essentially of two unconnected, insulated coils, called **primary terminal** and **secondary terminal,** wound around a hollow core of laminated iron (Fig. S-9).

Volt-ampere always enters into a primary terminal and exits from a secondary terminal.

The voltage appearing on each terminal has a direct proportion to the number of winding turns.

### Example 1

Assume 120 volts are entering a primary terminal and 240 volts are exiting from a secondary terminal.

a.   This is called **step-up transformer** and is indicated as 120/240-V.
b.   The number of winding turns are as follows:

For 120-V primary terminal, 500 turns

For 240-V secondary terminal, 1000 turns

### Example 2

Assume 240 volts are entering a primary terminal and 120 volts are exiting from a secondary terminal.

a.   This is called **step-down transformer** and is indicated as 240/120-V.
b.   The number of winding turns are as fcllows:

For 240-V primary terminal, 1000 turns

For 120-V secondary terminal, 500 turns

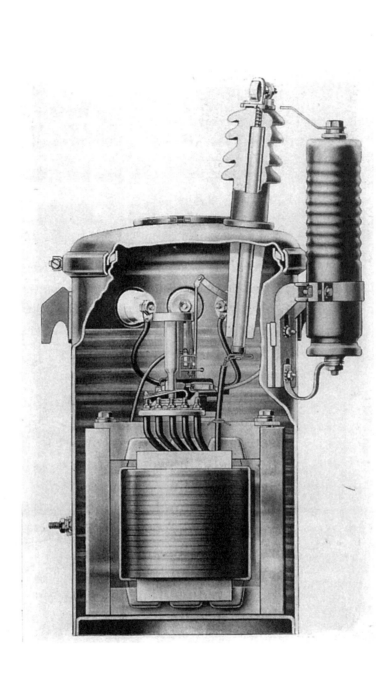

**Figure S-8    Cutaway View of Distribution Transformer**

## Example 3

Assume we have designed an electrical system for a building and the total load required is 150 amps at 120/240 V. The electric company's distribution line contains 2400 V.

a.  What are the amperes on the primary and secondary terminals?

b.  How can we specify the size of the transformer required for this building?

### Solution

a.  $VI = 150 \text{ A} \times 120 \text{ V} = 18,000 \text{ VA or } 18 \text{ kVA}$

$$\text{Primary terminal} = \frac{18,000 \text{ VA}}{2400 \text{ V}} = 7.5 \text{ A}$$

$$\text{Secondary terminal} = \frac{18,000 \text{ VA}}{120 \text{ V}} = 150 \text{ A}$$

b.  We need 18 kVA 2400/120–240-V transformer.


## S-15  HOW TO SPECIFY TRANSFORMERS

Transformers are specified as follows:

a.  **Type.**  There are two types dry (air cooled) and liquid filled. Up to 3000 kVA are commonly dry type.

b.  **Phase.**  Phases are given in S-8 to S-13 supra.

c.  **Voltage.**  Voltages are given in S-8 to S-13 supra.

d.  **kVA rating**

$$\text{kVA} = \text{kW}$$

**Transformer load is always designated by kVA rating.**

e.  **Sound level.**  Sound level maximum is 50 db (decibel).

f.  **Insulation class**

| Insulation type | Temperature avg. rise | Max. |
|---|---|---|
| **Organic** | 131°F | 302°F |
| **Mica, glass, resins** | 176°F | 1302°F |
| **Silicons** | 302°F | 428°F |

139

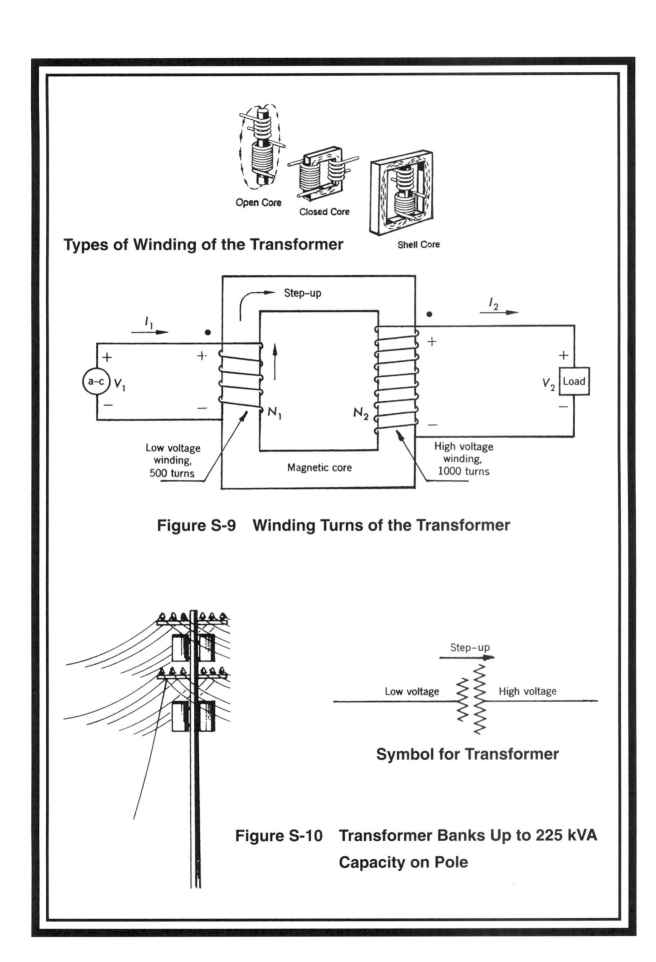

Types of Winding of the Transformer

Open Core

Closed Core

Shell Core

Step-up

$I_1$

$I_2$

a-c $V_1$

$V_2$ Load

Low voltage
winding,
500 turns

$N_1$

$N_2$

Magnetic core

High voltage
winding,
1000 turns

**Figure S-9   Winding Turns of the Transformer**

Step-up

Low voltage                High voltage

**Symbol for Transformer**

**Figure S-10   Transformer Banks Up to 225 kVA**
**Capacity on Pole**

*Example*

Using the building given in S-14, Example 3, specify the transformer.

*Solution*

    18 kVA, single-phase, 2400/120-240 V

    Air cooled, outdoor dry type with 302°C insulation and 176°F rise, with

    50-db max. sound level


**S-16**    **LOCATION OF TRANSFORMER**

For a residence or small building with a load requirement of approximately 18 kVA (150-A, 120/240-V), the utility company commonly installs one transformer to serve several buildings.

The transformer is located on the pole in **overhead service** (Fig. S-10), or it is located on the ground over a concrete plate as is shown in (Fig. S-11) for **underground service.**

For a large project with an approximate load requirement of GkVA (500-A, 277/480-V) or more, the transformer will be installed within the project. It may be located:

1.    **Outdoors**

2.    **Indoors**


**S-17**    **OUTDOOR TRANSFORMER**

It is a good practice to locate the transformer outside the building(s) if the required space is available for the following reasons:

1.    It saves valuable space within the building.

2.    Heat produced by a transformer does not create problems.

3.    Replacement and repairs will be handled with ease.

4.    It is less costly to install and maintain.


**S-18**    **INDOOR TRANSFORMER**

When a building(s) occupies the entire front and side yards, there is no choice except to locate the transformer inside the structure.

**Outdoor Transformer**  **Indoor Transformer**

**Figure S-11**

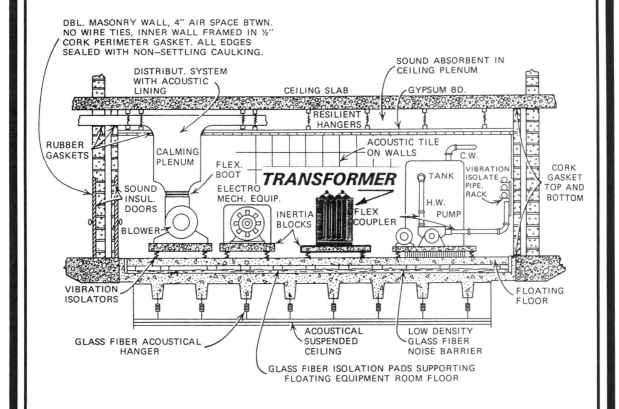

DBL. MASONRY WALL, 4" AIR SPACE BTWN.
NO WIRE TIES, INNER WALL FRAMED IN ½"
CORK PERIMETER GASKET. ALL EDGES
SEALED WITH NON—SETTLING CAULKING.

SOUND ABSORBENT IN
CEILING PLENUM

DISTRIBUT. SYSTEM
WITH ACOUSTIC
LINING

CEILING SLAB

GYPSUM BD.

RESILIENT
HANGERS

RUBBER
GASKETS

ACOUSTIC TILE
ON WALLS

C.W.

CALMING
PLENUM

FLEX.
BOOT

**TRANSFORMER**

TANK

VIBRATION
ISOLATE
PIPE.
RACK

CORK
GASKET
TOP AND
BOTTOM

SOUND
INSUL.
DOORS

ELECTRO
MECH. EQUIP.

H.W.
PUMP

BLOWER

INERTIA
BLOCKS

FLEX
COUPLER

VIBRATION
ISOLATORS

FLOATING
FLOOR

GLASS FIBER ACOUSTICAL
HANGER

ACOUSTICAL
SUSPENDED
CEILING

LOW DENSITY
GLASS FIBER
NOISE BARRIER

GLASS FIBER ISOLATION PADS SUPPORTING
FLOATING EQUIPMENT ROOM FLOOR

**Figure S-12   Soundproofing a Mechanical Equipment Room**

*U.S. Dept. of Housing and Urban Development*

**142**

The principles for installation of an indoor transformer are:

1. ***Noise***

    *a.* Specify transformer with maximum sound leve of 50 db (decibel).

    *b.* Locate transformer in the least valuable space in the building.

    *c.* ***Do not*** place the transformer in the quiet area of the structure.

    *d.* Treat the transformer room with acoustical materials in order to increase the sound absorption (Fig. S-12).

2. ***Vibration***

    A transformer vibrates and will transfer this vibration to other parts of the building. To correct this situation:

    *a.* Use vibration isolators under a concrete pad to receive the transformer (Fig. S-12).

    *b.* Use a flexible concuit for a load entering and leaving the transformer (Fig. S-12).

3. ***Heat***

    Heat generated from transformer at ful load is approximately **1.5 percent** of the transformer rating.

    ***Example***

    A transformer is rated at 800 kVA. What is the total heat produced by this transformer?

    ***Solution***

    800 kVA × 1000 = 800,000 $VI$

    800,000 $VI$ × 0.015 = 12,000 $VI$ = W

    1 W produces 3.41 Btu (please see E-36).

    12,000 W × 3.41 Btu/W = 40,920 Btu heat produced.

    In order to avoid overheating the transformer:

    *a.* Provide approximately **3 sq. ft. per kVA** capacity of transformer for the size of adjustable louver installed in an exterior wall or door adjacent to the transformer.

    *b.* Use **4 sq. in. per kVA** if switchgear is used.

### Example

A transformer is rated at 800 kVA with switchgear. What is the size of adjustable louver required on the exterior wall adjacent to the transformer?

### Solution

800 kVA × 4 in/kVA = 3200 sq. in.

Adjustable louver is 3'-0" × 7'-6", or two 3'-0" × 3'-9" louvers

4.  See regulations for installation of a transformer in the electrical code used in the area where your building will be constructed.

# ELECTRICAL MOTORS

### S-19    ELECTRICAL MOTORS

**Motors are used to change the electrical energy to mechanical energy.**

**Electrical energy ────➤ Motor ────➤ Mechanical energy**

A motor consists of a *stator* mounted in a frame and a cylindrical *rotor,* also known as *armature.*

Held by bearings on both sides of the frame, it is free to rotate inside of the stator (Fig. S-13).

Laminated copper wires are wound around or through laminated iron pieces of the stator, called *field winding,* and also wound around the rotor, called *rotor winding,* or armature winding. In order to minimize an eddy-current loss of the motors, the armature iron is made of thin laminations.

When current is applied to the copper wire of the stator, it creates a magnetic field, and current passing through the rotor responds to the forces of the magnetic field and starts to rotate.

When **DC** is supplied to both the stator and the rotor, the motor is called *DC motor* (Figs. S-14 and S-15).

**DC ────➤ Stator**
**DC ────➤ Rotor** ────➤ **DC motor**

When **DC** is applied to the stator and **AC** is supplied to the rotor, the motor is called *synchronous motor* (Figs. S-16 and S-17).

**DC ────➤ Stator**
**AC ────➤ Rotor** ────➤ **Synchronous motor**

When AC is supplied directly to the stator and by induction to the rotor, the motor is called *induction motor* (Figs. S-18 and S-19).

**AC ────➤ Stator ────➤ Rotor ────➤ Induction motor**

The stator winding of an induction motor is similar to the synchronous motor.

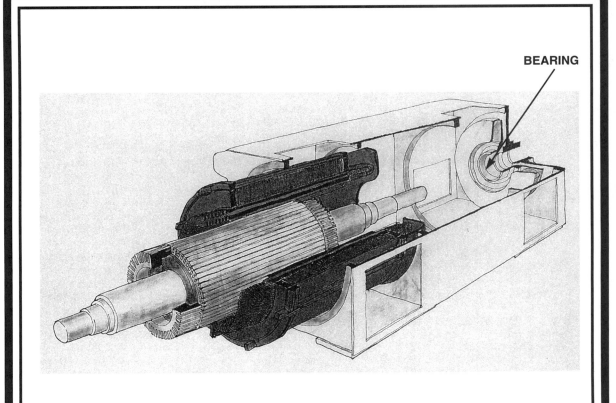

BEARING

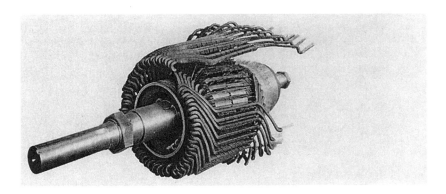

*DC generator or motor armature in process of being wound. One side of each coil is placed in the bottom of a slot; the other side is placed in the top of a slot.*

## Figure S-13    Motor Armature

## S-20 USE OF MOTORS

Motors are used to provide mechanical energy for operating mechanical systems and appliances within and around the buldings. They include, but are not limited to:

a. Fans

b. Blowers

c. Pumps

d. Compressors

e. Vertical transportation


## S-21 MOTORS CLASSIFICATION

Motors are classified according to:

a. **Size**

In horsepower **(HP = 740 W)**

b. **Type**

1. Application

2. Enclosures (open drip-proof and totally enclosed fan-cooled)

c. **Electrical characteristics "kVA"**

Also max. starting current per HP

d. **Frame**

Motor's dimensions


## S-22 TYPES OF MOTORS

The motors used for mechanical systems, vertical transportation, appliances, and devices in building(s) are manufactured from $\frac{1}{10}$ **HP, 74.6 W,** to as high as **20 HP, 14,920 W** and larger.

There are seven types of motors:

1. **Split-phase motors**

2. **Capacitor motors**

3. **Repulsion-type motors**

4. **Polyphase motors**

5. **Shaded-pole motors**

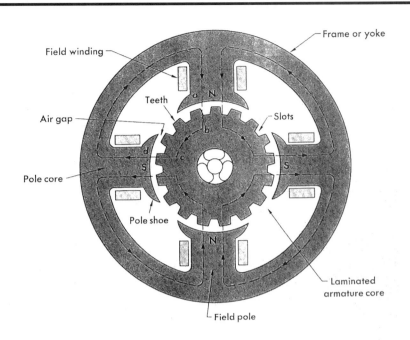

**Figure S-14   Diagrammatic Sketch of DC Machine**

**Figure S-15   Cutaway View of Typical DC Motor**

6. **Universal motors**

7. **Direct-current motors**

**The National Electric Manufacturers Association (NEMA)** has established guidelines for the design of electrical motors, which are followed by many manufacturers.

### S-23  SPLIT-PHASE MOTORS

They are often called *squirrel-cage induction motors.*

**Squirrel-cage induction motors consist of:**

*a.*  **Housing.**

*b.*  **A laminated iron core** *stator* with embedded wind ng constitutes the inside of the housing.

*c.*  *Rotor* consists of copper bars set in iron core slots which are connected to each other with copper rings.

A centrifugal *switch* is located inside the housing to open the circuit to the starting winding when the motor reaches the designed speed.

Squirrel-cage induction motors are manufactured in four different types based on **NEMA** design guidelines, two of which are commonly used for mechanical systems in building(s):

    **Type B**   *Standard design.*   It has high power factor **(PF)** and high efficiency **(E).** It is used for pumps, fans, and blowers, etc.

    **Type C**   *High starting torque.*   It has an average power factor and average efficiency. It is used for equipment that starts under load, such as conveyors, compressors, etc.

### S-24  CAPACITOR MOTORS

The design of this motor is similar to the squirrel-cage motor except a *capacitor* is placed in a series before the centrifugal switch and starting windings.

The capacitor produces higher starting torque with lower starting current.

Capacitor motor sizes are from under **1 HP** to approximately **12 HP,** and operate with single-phase **AC.**

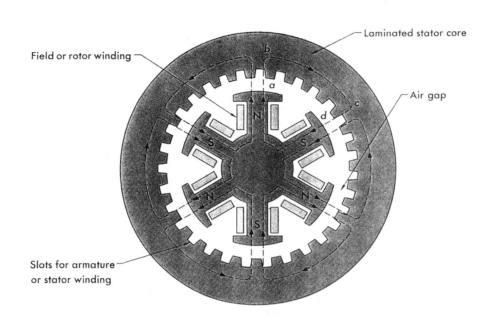

**Figure S-16   Diagrammatic Sketch of Salient-Pole Synchronous Machine**

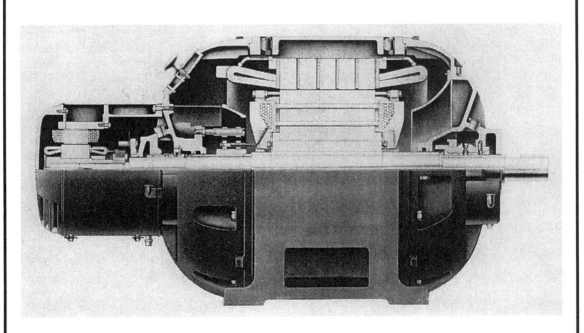

*At the left end is a shaft-mounted DC generator to furnish excitation to the motor field.*

**Figure S-17   Cutaway View of High-Speed Synchronous Machine**

There are two types of capacitor motors:

1.   **Capacitor-start motors**
2.   **Capacitor start-and-run motors**

They are both smooth and quiet motors and require less maintenance; however, they have low starting torque.

They are both commonly used for fans, blowers, and compressors used for refrigerators and heating, ventilating, and air-conditioning systems (HVAC).

S-25   **REPULSION-TYPE MOTORS**

This motor is manufactured only as a brush-riding type, and it starts and runs on the repulsion principle.

It has a variable speed, high starting torque, and, by adjusting the brush holder, the speed can be decreased and rotation also can be reversed.

It is available in sizes from ⅒ **HP, 74.6 W,** to about **20 HP, 14,920 W.**

S-26   **POLYPHASE MOTORS**

The design of these motors is identical to the design of split-phase motors without a centrifugal switch.

The stator is not connected to the rotor electrically; and the magnetic field created by the stator induces voltage to the rotor similar to that of the transformer. For this reason the stator is called *primary* and the rotor is called *secondary.*

Polyphase motors are:

*a.*   **Three-phase AC motors**
*b.*   **Extremely efficient and economical**
*c.*   **Designed for all standard voltage**
*d.*   **Known for their good, constant speed**
*e.*   **Available with a variety of torque characteristics**
*f.*   **Designed for normal and high starting current**

They are available in **a great variety of sizes (HP)** and are used for many applications, especially in commercial and industrial structures.

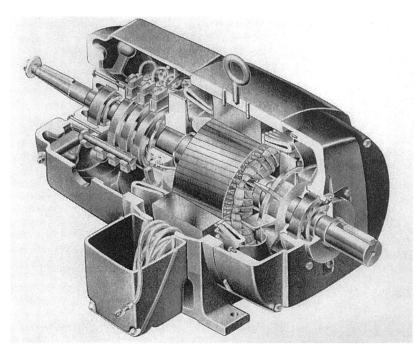

**Figure S-18    Cutaway View of 3-Phase Induction Motor with a Wound Rotor and Slip Rings Connected to the 3-Phase Rotor Winding**

*General Electric Co.*

**Figure S-19    Cutaway View of an Induction Motor with Squirrel-Cage Rotor**

*Westinghouse Electric Corp.*

## S-27 SHADED-POLE MOTORS

This motor is a single-phase induction motor with a constant speed and low starting torque.

**It is used for small fans, clocks, relays, and other small electrical devices.**

## S-28 UNIVERSAL MOTORS

This motor consists of field windings on the stator and windings on the rotor. The carbon brushes are placed in such a way that they allow proper contact with the commutator.

The current flows through the stator and rotor in a series, and the magnetic field created by the stator causes the rotor to rotate.

This motor operates with a single-phase 120-V AC or 240-V DC.

**It is used for small fans, small kitchen appliances, hair dryers, shavers, etc.**

## S-29 DIRECT-CURRENT MOTORS

This motor utilizes only direct current.

It consists of a frame called **york,** field winding **stator**, and a laminated armature **rotor**.

The typical armature is shown in Fig. S-13, a schematic diagram of DC motors, Fig. S-14, and cutaway view, Fig. S-15.

Direct-current motors are of three types:

1. **Series motor**
2. **Compound motor**
3. **Standard shunt motor**

## S-30 DC SERIES MOTOR

In this motor the field coils are in a series with a rotor and the line.

A series motor must always have an adequate load in order to maintain a proper speed; therefore, the speed varies with the change in the load.

## S-31    DC COMPOUND MOTOR

The design of this motor is similar to the design of the compound generator.

The compound motor features are a combination of both standard shunt motors and series motors.

It has a greater starting torque and variable speed under changing loads than the series motor.


## S-32    DC STANDARD SHUNT MOTOR

The design of the standard shunt motor is similar to the design of the shunt generator. As the load on the shunt motor increases, the rotor tends to slow down (as in series and compound motors). At this instance, the voltage is reduced automatically, allowing more current to flow into the rotor and causing the rotor to maintain a constant speed.

The features of the standard shunt motor are:

*a.*    **It produces up to two times the full load torque at the start.**

*b.*    **The speed rapidly increases to full speed.**

*c.*    **It practically has a constant speed under variable loads.**

*d.*    **The speed can be varied from 75 to 100 percent.**

*e.*    **The motor can be designed to have varied speeds from 30 to 100 percent.**

For these reasons, **the standard shunt motor is used to operate the vertical transportation systems.**

Because the electric power available in the building is alternating current, an **AC motor** is coupled to a **DC generator to provide direct current.** This direct current is supplied to **a DC motor** of a gearless traction machine or geared traction machine of the vertical transportation system.

## S-33   CONTROLS USED TO OPERATE MOTORS

The operation of any motor depends directly on one or many control devices according to its function.

The controls for a motor may include a few or all of the following devices and more:

*a.* **Overcurrent protection.**   Fuses or circuitbreakers.

*b.* **Switch(es).**   For starting and stopping the motor.

*c.* **Changing speed.**

*d.* **Reversing.**

*e.* **Overload protection.**   To stop the motor if the load applied exceeds the design load.

*f.* **Sequence control.**   A device that acts in time order or in series.

*g.* **Jogging.**   To stop the motor when it starts jogging.

*h.* **Plugging.**   Breaking an induction motor by reversing the phase sequence of the power.

*i.* **Pilot light induction.**   A white, red, or green light indicating the condition of the operation.


## S-34   DESIGN CRITERIA

Local codes and **NEC** have rules governing motors, motor circuits, and controllers, etc.

Check the code in your local area before preparing the construction documents. The following is for your information only:

Motor(s) should be located in an area that allows adequate ventilation to cool the motor(s) and allow easy maintenance.

If motors are operating with more than 50 volts, their exposed live parts terminals must be grounded.

Motors should be located in a room enclosure accessible only to a qualified person.

Motors with current requirements of approximately 10 A or more need their own branch circuit with fuses or circuit breakers.

# CONTENTS

# Part 4

## ELECTRICAL CONDUCTORS (WIRING)

# C

### Electrical Conductors (Wiring)

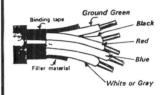

*CABLE*

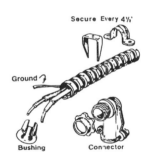

158

# ELECTRICAL CONDUCTORS (WIRING)

## C-1 HISTORY AT A GLANCE

Use of wire is dated to 3000 B.C. The metal was hammered into sheets, cut in strips, and again shaped with a hammer to form wire.

The drawing of the wire started in Europe in the 13th century. The metal was drawn in a series of holes, each smaller than the one preceding, until the wire was drawn through the hole with the desired diameter. The plates with such holes were and still are known as *drawplates* or *dies.*

*Ichabod Crane,* an American blacksmith, in 1831 began to manufacture draw wire by using the water wheel as mechanical power. He invented many devices in draw wire, some of which form the basis for modern wire drawing machinery.

With the invention of the **telegraph, telephone,** and **electricity** in the late 1800s, manufacturing of wire became one of the largest industries in the 20th century.

## C-2 ELECTRICAL CONDUCTORS (WIRING)

Conductors (wiring) are used to carry the current at certain voltage through the electrical system. (Please see E-4, Electricity Moving in Conductors.)

In the United States there are two categories of conductors, based on their diameter and **DC** resistance.

1. **AWG**  American Wire Gauge
2. **MCM**  Thousand "M" Circular Mils

"**M**" is the Roman numeral for thousand.

## C-3 AMERICAN WIRE GAUGE AWG

The pioneers who established the wiring numbers for **AWG** probably had good reasons for creating the following systems.

We may not agree or like the numbering system; however, we have to respect and be grateful for their efforts.

| SIZE | AREA | DIAMETER (in inch) | |
|---|---|---|---|
| | (C.M.) | SOLID | STRANDED |
| _AGW_ | | | |
| 16 | 2580 | 0.0508 | ----- |
| | _WIRE_ | | |
| 14 | 4109 | 0.0641 | ----- |
| 12 | 6530 | 0.0808 | ----- |
| 10 | 10,380 | 0.1019 | ----- |
| 8 | 16,510 | 0.1285 | ----- |
| | _CABLE_ | | |
| 6 | 26,240 | 0.162 | 0.184 |
| 4 | 41,740 | 0.204 | 0.232 |
| 2 | 66,360 | 0.258 | 0.292 |
| 1 | 83,690 | 0.289 | 0.332 |
| 0 (1/0) | 105,600 | 0.325 | 0.373 |
| 00 (2/0) | 133,100 | 0.365 | 0.418 |
| 000 (3/0) | 167,800 | 0.410 | 0.470 |
| 0000 (4/0) | 211,600 | 0.460 | 0.528 |
| _MCM_ | | | |
| 250 MCM | 250,000 | 0.500 | 0.575 |
| 300 MCM | 300,000 | 0.548 | 0.630 |
| 400 MCM | 400,000 | 0.632 | 0.728 |
| 500 MCM | 500,000 | 0.707 | 0.813 |

**Figure C-1   Physical Properties of Conductors**

Extracted from the National Electrical Code

AWG is preceded by a number which indicates the size of wire. For example, "**14 AWG**" means "**American Wire Gauge number 14.**"

AWG numbers run in reverse order to the size of the conductor.

AWG used for electrical systems starts with number **14 AWG** (the smallest-size wire, with a diameter of 0.0641 in.), then 12 AWG, 10, 8, 6, 4, 2, 1, 0, 00, 000, and **0000 AWG** (the largest size cable, with a diameter of 0.46 in.) (Fig. C-1).

*Note:* Smaller-size **AWG** are available (16 AWG, 18, 20, 22 AWG, etc.), but they are not permitted to be used in electrical circuits.

There are two categories in **AWG** systems:

a.   *Wire*

A single conductor number 14 **AWG, 12 AWG, 10 AWG,** and **8 AWG** is called *wire*.

b.   *Cable*

1.   **A single conductor number 6 AWG to 0000 AWG is called** *cable.*
2.   **Also, two or more wires assembled in a single jacket are called** *cable.*
3.   **Conductors 6 AWG and larger are** *stranded.*

**For "AWG zeros" the following are also used:**

0 AWG = 1/0 AWG

00 AWG = 2/0 AWG

000 AWG = 3/0 AWG

0000 AWG = 4/0 AWG

**In the electrical trade the following pronunciations are used:**

| | |
|---|---|
| 1/0 AWG | One aught or one naught |
| 2/0 AWG | Two aught or two naught |
| 3/0 AWG | Three aught or three naught |
| 4/0 AWG | Four aught or four naught |

## C-4    MCM THOUSAND CIRCULAR MILS

Electrical power ($VI$ = W) requirements for a large project require conductors with a larger diameter than 4/0 AWG.

For this reason a system of measurement was devised to accommodate this goal. It is called "MCM."

**M = Thousand**

**C = Circular**

**M = Mils**

Circular mil is an artificial measurement used to represent the diameter of a conductor; it begins with 500 mils.

500 mils (M) = 0.5 in. diameter (D)

**"MCM"**

**M = 1000      CM = $D^2$**

Therefore,      $$\text{MCM} = \frac{CM}{1000} = \frac{D^2}{1000}$$

For 500 Mils,      $$\text{MCM} = \frac{(500)^2}{1000} = \frac{250,000}{1000} = 250$$

**A conductor with a diameter of 0.5 in. or 500 mils is called "250 MCM."**

**MCM used for electrical systems start with 250 MCM** (the smallest size), **then 300 MCM, 400 MCM, and 500 MCM** (the largest size). These are sufficient sizes for our needs in structures, and the figures given in this book stop with conductor sizes up to **500 MCM** (Fig. C-1).

However, conductors with larger diameters are available and are used for projects requiring excessive amounts of power. They include **600,** 700, 800, 900, **1000 MCM** (diameter 1 inch), 1250, 1500, 1750 and **2000 MCM** (diameter 2 inches).


## C-5    METALS USED FOR CONDUCTORS

There are two types of metals used for manufacturing conductors:

1.    **Copper**

2.    **Aluminum**

## C-6 COPPER CONDUCTORS

Copper was one of the first metals known to humans, as long ago as 500 B.C.

Copper is ductile, malleable, and an excellent conductor of heat and electricity.

It is harder than zinc and softer than iron.

Approximately 50 percent of its total output is used for manufacturing of electrical apparatus and conductors. (Please see E-4, Electricity Moving in Conductors.)

All figures given in this book are for copper conductors, except a few which refer to both copper and aluminum.

## C-7 ALUMINUM CONDUCTORS

Various compounds of aluminum were used in Roman times and were called "alum."

In the 19th century, aluminum was recognized as a metal, with alumina as its oxide.

In 1886 the method of producing aluminum was discovered in the United States by C. M. Hall.

An extensive amount of electric power is required to produce aluminum.

The development of low-cost electricity by hydroelectric power greatly aided the expansion of the aluminum industry.

Aluminum is the formation of many materials such as feldspar, mica, alum, cryolite, clay bauxite, and several forms of aluminum oxide (alumina).

Aluminum's toughness, fairly high strength, and light weight account for its use.

Because of the rise in the price of copper in the last 25 years, a small percentage of conductors are made with aluminum.

**Advantages and disadvantages of aluminum conductors are as follows:**

1. They are lighter in weight than copper conductors.
2. They cost less than copper conductors.
3. Aluminum conductors have approximately 80 percent of the conductivity of copper conductors.

| *Trade Name* | *Type Letter* | *Application Provisions* | *Insulation* |
|---|---|---|---|
| Moisture- and heat-resistant rubber | RHW-2 | Dry and wet locations | Moisture- and heat-resistant rubber |
| Silicone-asbestos | SA | Dry and damp locations For special application | Silicone rubber |
| Synthetic heat-resistant | SIS | Switchboard wiring only | Heat- resistant rubber |
| Thermoplastic and asbestos | TA | Switchboard wiring only | Thermoplastic and asbestos |
| Thermoplastic | TBS | Switchboard wiring only | Thermoplastic and fibrous outer braid |
| Moisture- and heat-resistant thermoplastic | THWN | Dry and wet locations | Flame-retardant, moisture- and heat-resistant thermoplastic |
| Moisture-resistant thermoplastic | TW | Dry and wet location | Flame-retardant, moisture-resistant thermoplastic |

**Figure C-2 Conductors, Applications, and Insulations**

Extracted from the National Electrical Code

4.  Copper conductors up to approximately 4 AWG require smaller conduits; therefore, the cost is generally lower than aluminum conductors within a conduit.

5.  If an aluminum conductor is not properly installed it will oxidize under excessive pressure, and heat generated from the power causes the joints to loosen, which may cause incendiary effects.

6.  An aluminum conductor can create electrical problems if installed improperly.

Because of some unfortunate incidents, some local codes have banned the use of aluminum conductors in branch circuitry.

## C-8   MAXIMUM VOLTAGE DROP IN CONDUCTORS

Conductors have a voltage rating (300 V, 600 V, etc.).

The voltage applied to a conductor should be equal to or less than the voltage rating specified.

**Conductors should be designed for:**

*a.*   **Voltage drop not to exceed 3 percent of power for lighting, heating, or a combination of lighting and heating**

*b.*   **Voltage drop for branch circuits, conductors, and feeders which combined should not exceed 5 percent**

**To obtain the percentage of a voltage drop:**

$$VD = 2\,K \times L \times \frac{I}{CM}$$

$$\frac{VD}{\textbf{Voltage in circuit}} = \textbf{\% of voltage drop}$$

where   $VD$ = Voltage drop in circuit

$K$ = **Resistivity of a conductor**

**Use (11) for copper conductor**

**and (18) for aluminum conductor**

$L$ = Length of conductor

$I$ = Current of circuit

$CM$ = Area of conductor in circular mils given in Fig. C-1

$V$ = Voltage in circuit

| Trade Name | Type Letter | Application Provisions | Insulation |
|---|---|---|---|
| Perfluoroalkoxy | PFA | Dry and damp locations. Dry location-special applications. | Perfluoroalkoxy |
| Perfluoroalkoxy | PFAH | Dry locations only. Only for leads within apparatus or within raceways connected to apparatus. (nickel or nickel-coated copper only). | Perfluoroalkoxy |
| Heat-resistant rubber | RH | Dry and damp locations. | |
| Heat-resistant rubber | RHH | Dry and damp locations. | Heat-resistant rubber |
| Moisture- and heat-resistant rubber | RHW | Dry and wet locations. For over 2000 volts insulation shall be ozone-resistant. | Moisture- and heat-resistant rubber |

**Figure C-3  Conductors, Applications, and Insulations**

Extracted from the National Electrical Code

## Example

A 12 AWG copper conductor is 35 feet long and carries 20 amperes in a circuit with 120 volts. What is the voltage drop?

## Solution

$K = 11$ for copper conductor

$CM = 6530$ from Fig. C-1

$$VD = 2 \times 11 \times 35 \times \frac{20}{6530} = 2.3562 \text{ V}$$

Percentage of voltage drop $= \dfrac{2.3562}{120 \text{ V}} = 0.0196\%$

This is well below the 3 percent allowable.

## C-9  CONDUCTOR INSULATION

The conductor acts as a guide for the conduction of electrical power from one point to another point and must be insulated to prevent the escape or leakage of the current.

The types of insulation used are:

*a.*   **Moisture- and heat-resistant rubber**

*b.*   **Heat-resistant rubber**

*c.*   **Silicone rubber**

*d.*   **Thermoplastic and asbestos**

*e.*   **Thermoplastic**

*f.*   **Flame resistant, moisture- and heat-resistant thermoplastic**

*g.*   **Moisture-resistant**

*h.*   **Perfluoroalkoxy**

The list of recommended insulation to be used for different applications is given in Figs. C-2 and C-3.

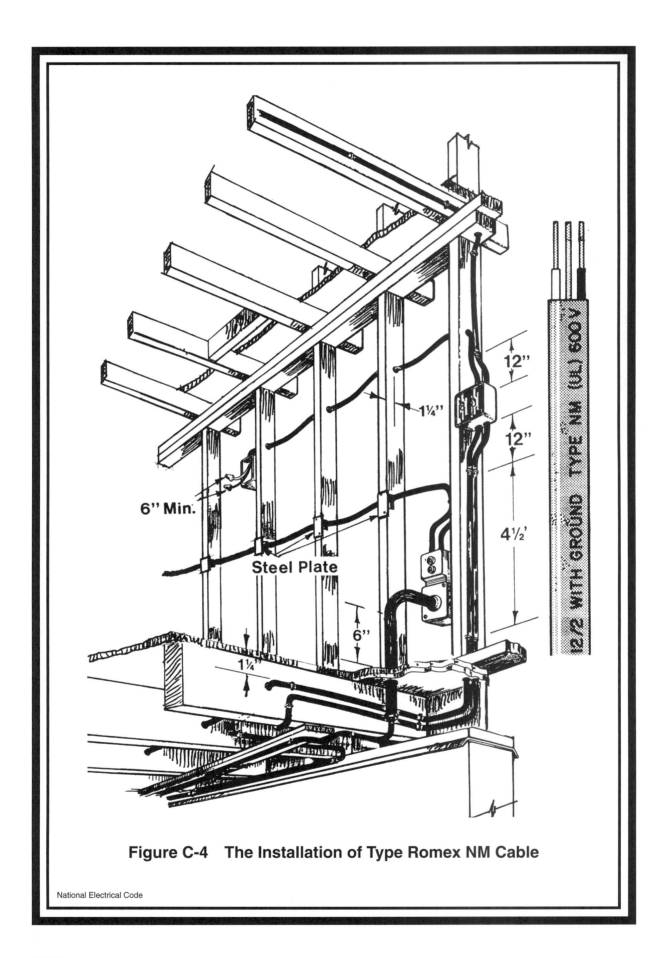

12"

12"

1¼"

6" Min.

Steel Plate

4½'

6"

1¼"

12/2 WITH GROUND   TYPE NM (UL) 600 V

**Figure C-4   The Installation of Type Romex NM Cable**

National Electrical Code

## C-10 WIRING SYSTEMS

There are two types of applications in electrical wiring systems:

1. **Wiring to provide power to operate**

   a. Lighting and outlets (receptacles)

   b. Equipment

2. **Control wiring to operate equipment**

Wiring systems may be:

1. **Exposed wiring,** where wires or feeders are installed on the surface of walls, ceilings, etc.

2. **Concealed wiring,** where wires or feeders are concealed inside the walls, ceiling, etc.

## C-11 CABLE SYSTEMS

Several types of cable systems are used in wiring the building:

1. *Nonmetallic sheathed cable*

2. *Flexible armored cable*

3. *Underground feeder cable*

4. *Service-entrance cable*

5. *Mineral-insulated metal-sheathed cable*

## C-12 NONMETALLIC SHEATHED CABLE

This cable is commonly known as *romex.*

Romex cables type **NM** are manufactured in two or three wires with conductors varying in sizes.

Uncoated copper conductors with **PVC** insulation are color-coded with a bare ground wire (Figs. C-4 and C-8b).

The jacket is made of rubber, plastic, or fiber.

This type of cable may be concealed or exposed.

Some codes may not allow the use of this type of cable or may restrict the use for residential and other small structures not exceeding three stories.

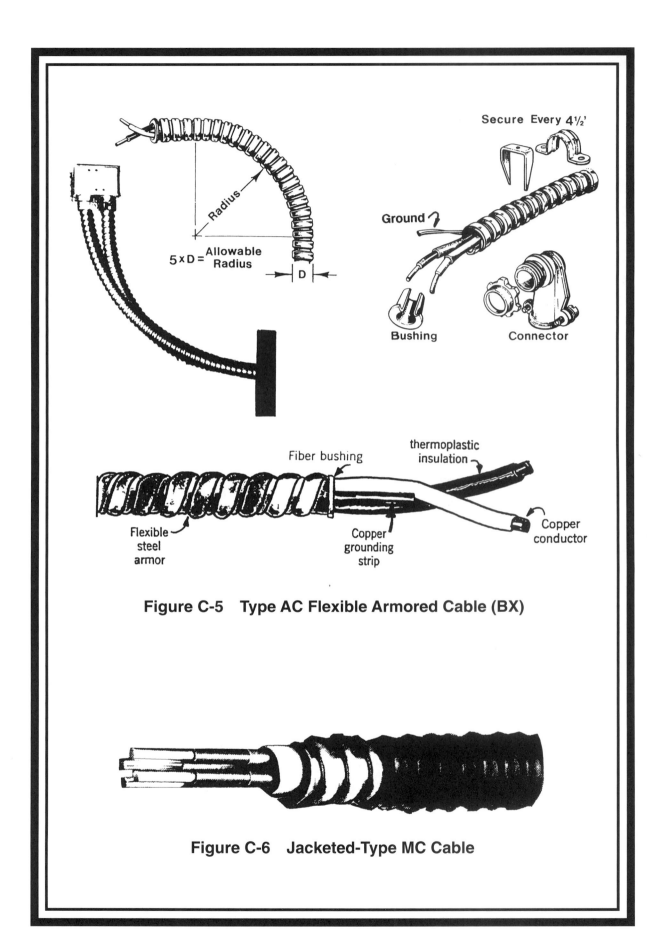

Secure Every 4½'

Ground

Bushing

Connector

Radius

$5 \times D = \frac{\text{Allowable}}{\text{Radius}}$

D

thermoplastic
insulation

Fiber bushing

Flexible
steel
armor

Copper
grounding
strip

Copper
conductor

**Figure C-5   Type AC Flexible Armored Cable (BX)**

**Figure C-6   Jacketed-Type MC Cable**

## C-13   FLEXIBLE ARMORED CABLE

This cable is commonly referred to as *BX cable.*

It is manufactured in two-, three-, and four-wire assemblies with insulated wires wrapped with a spiral-wound interlocking steel jacket.

There are two types of BX cables:

1. *BX cable type "AC."*   This cable is restricted to dry locations (Fig. C-5).
2. *BX cable type "MC."*   This cable has an additional **PVC jacket,** and can be used in wet locations (Fig. C-6).

## C-14   UNDERGROUND FEEDER CABLE

Underground-feeder-type **UF** cable can be used for direct burial.

When it is used above the ground and exposed to the sun, the jacket must be sun-resistant.

## C-15   SERVICE-ENTRANCE CABLE

For service-entrance cable, type **SE** may be used.

It must be installed according to **NEC** or the local code.

This cable can be used in the interior wiring system if the cable is insulated with thermoplastic or rubber.

## C-16   MINERAL-INSULATED METAL-SHEATHED CABLE

This cable is type **MI.**

It is manufactured with one or more conductors.

It is insulated with compressed mineral insulation, and enclosed in a gas-tight and liquid-tight continuous copper sheath (Fig. C-7).

This type of cable can be used for a service feeder or circuit branch.

It can be used in a dry or wet location, indoors or outdoors.

It can also be used in hazardous areas.

**171**

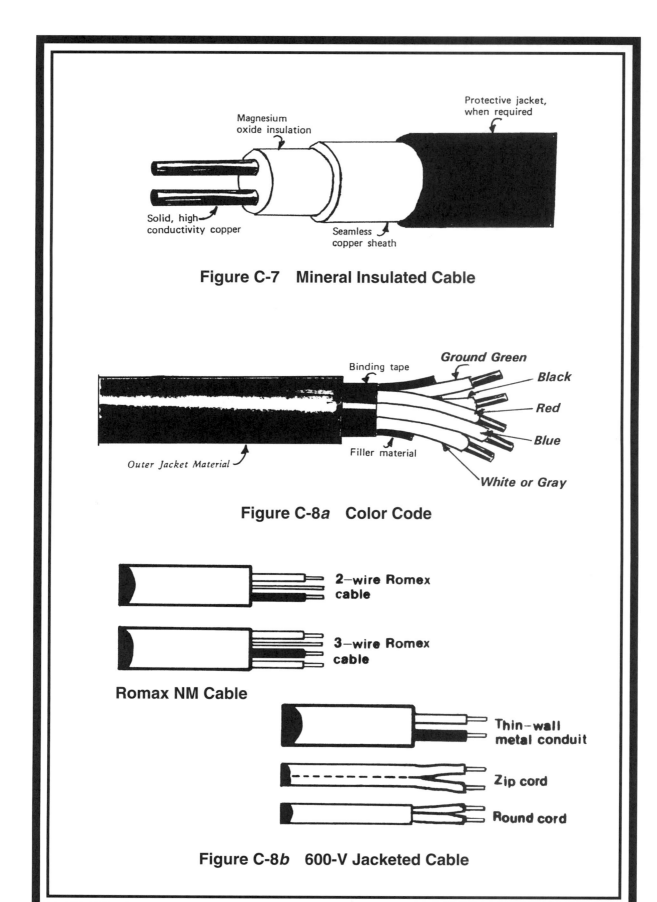

Figure C-7   Mineral Insulated Cable

Figure C-8*a*   Color Code

**Romax NM Cable**

Figure C-8*b*   600-V Jacketed Cable

## C-17 WIRING SYSTEMS COLOR CODE

Conductors' insulations are color-coded during the manufacturing as follows:

| | |
|---|---|
| **Phase A or (1)** | **black** |
| **Phase B or (2)** | **red** |
| **Phase C or (3)** | **blue** |
| **Neutral** | **white or gray** |
| **Ground** | **bare or green when insulated** |

Fig. C-8*a* shows the coloring code for a 600-V jacketed cable. All other cables have similar color codes.

## C-18 RACEWAYS

Any channels, pipes, or conduits and ducts which are designed especially to house wires or cables are called *raceways.*

They may be classified in the following categories:

*a.* **Underground raceways**

*b.* **Exposed and concealed raceways**

## C-19 UNDERGROUND RACEWAYS

Underground raceways which have resistance to corrosion are directly buried.

They are commonly manufactured in two types:

1. **Insulated impregnated fiber**
2. **Insulated cement-asbestos**

Underground raceways are normally installed in place first, and wires and/or cables are placed inside the raceway at a later time.

173

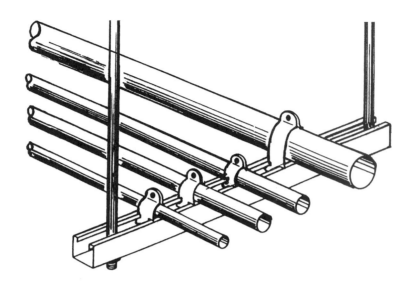

**Installation of Electrical Metallic Conduit (EMT)**

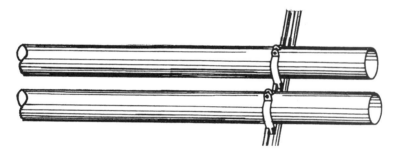

**Rigid PVC (Plastic) Conduit**

**Electrical Metallic Tubings with Fittings**

**Flexible Metal Conduit**

**Figure C-9**

174

## C-20    EXPOSED AND CONCEALED RACEWAYS

There are several types of raceways which are used in all types of buildings, either exposed or concealed, some of which are:

*a.* **Pipe or conduit**

*b.* **Wireway**

*c.* **Ceiling raceways**

*d.* **Floor raceways**


## C-21    PIPE OR CONDUIT RACEWAYS

The most common types of raceways are pipes or conduits. They can be used exposed or concealed.

After the pipe or conduit is installed the conductors are pulled into the conduit.

In general, there are four types (Fig. C-9):

1. **Metallic conduit**

2. **Aluminum conduit**

3. **Flexible metallic conduit**

4. **Nonmetallic conduit**


## C-22    METALLIC CONDUIT

The electrical codes and NEC in many cases require that all or a part of electrical conductors be enclosed in a rigid metallic conduit.

All rigid metallic conduits and their fittings are required to be corrosion-resistant.

Steel conduits are manufactured as follows:

*a.* **Hot-dip galvanized** (Fig. C-10)

*b.* **Enameled** (Fig. C-11)

*c.* **Sheranized** (coated with zinc) (Fig. C-12)

*d.* **Plastic-covered** (Fig. C-13)

# CONDUITS

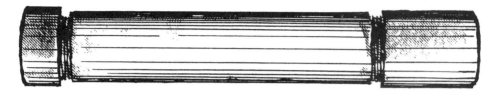

**Figure C-10   Galvanized, Heavy Wall, Rigid**

**Figure C-11    Black Enameled**

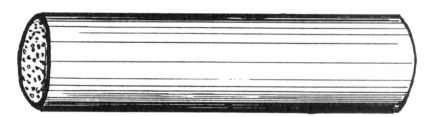

**Figure C-12    EMT Thin Wall**

**Figure C-13    Plastic-Coated Conduit for Use in Highly
Corrosive Atmospheres**

Metallic conduits are made and classified in three different wall thicknesses:

1. **Heavy-wall steel conduit**

   Known as *Rigid steel conduit* **"RS"**

   Wall thickness 0.113 inch

2. *Intermediate metal conduit* **"IMC"**

   Wall thickness 0.071 inch

3. *Electric metallic tubing* **"EMT"**

   Known as *thin-wall conduit*

   Wall thickness 0.05 inch

**Why metallic conduits are used:**

a. They protect the conductors from the hazard of corrosion and abuses.

b. Shock hazards can be avoided.

c. They provide a shield against fire hazards in the case conductors are or become over-heated.

d. They are a good support for conductors.

e. They provide a ground path for the wiring system.

## C-23  ALUMINUM CONDUIT

Aluminum conduits can be used in place of steel conduits in all locations. Check your local electric code.

**Aluminum conduits' advantages over steel conduits are:**

a. Lighter than steel concuits (Fig. C-14)

b. Less costly than steel conduits

c. Less labor costs than steel conduits

d. Better corrosion resistance

e. Lower voltage drop

f. Do not need painting

g. Nonsparking and nonmagnetic

Aluminum conduits should not be used underground because of corrosion.

If aluminum conductors are used in concrete, the concrete mix should be designed to prevent possible cracking.

**177**

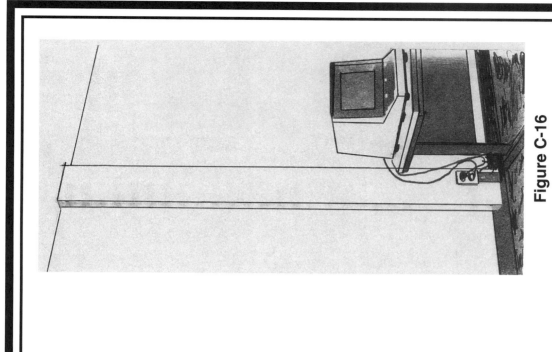

Figure C-16

## COMPARISON OF WEIGHT BETWEEN ALUMINUM (AL) AND STEEL (RS), (IMC), (EMT) CONDUCTORS

| Nominal size in inch | Weight of 10 ft. lengths in pounds | | | |
|---|---|---|---|---|
| | RS | IMC | EMT | AL |
| ½ | 7.9 | 5.7 | 2.9 | 2.7 |
| 1½ | 24.9 | 17.6 | 11.0 | 8.6 |
| 2½ | 52.7 | 39.3 | 20.5 | 18.3 |
| 3½ | 83.1 | 56.1 | 32.5 | 28.8 |

### Figure C-14

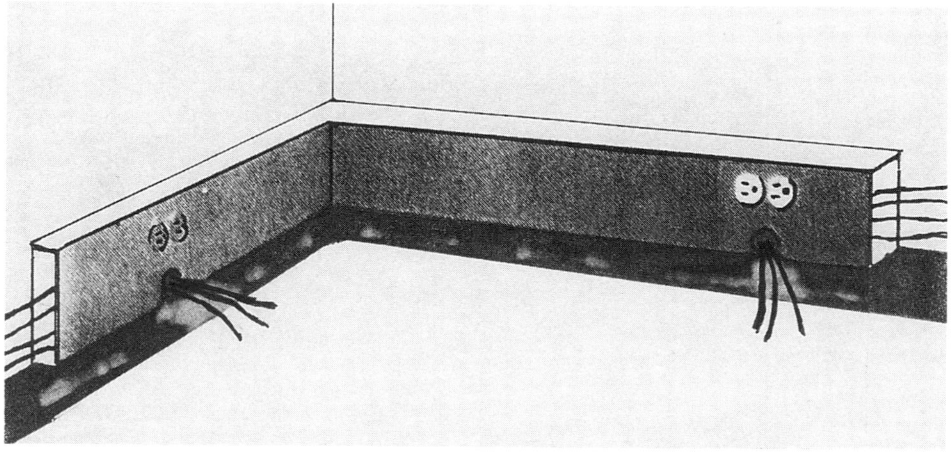

Figure C-15   SURFACE WIREWAY

178

## C-24 FLEXIBLE METAL CONDUIT

A flexible metal conduit is a halo spirally wound interlocked armer or raceway called **greenfield.**

It is manufactured with galvanized steel or aluminum (Fig. C-9).

**It is used in an electrical wiring system for:**

a. **Wiring connections to electrical equipment which vibrate** (motors, transformers, etc.)

b. **Where it is difficult to install a rigid conduit**

When it is covered with liquid-tight plastic jacket called **sealtite,** it can be used outside of the structure or inside in the moist and wet areas.

## C-25 NONMETALLIC CONDUIT

The nonmetallic conduits are classified as follows:

a. **Rigid polyvinyl chloride "PVC"**

b. **High-density polyethylene "HDPE"**

c. **Asbestos-cement (please see C-19)**

Nonmetallic conduits can be used without restrictions:

a. **In any nonhazardous structures**

b. **Both inside and outside of buildings**

**PVC** is less costly in material and labor than metallic or aluminum conduits.

**The required characteristics:**

| | |
|---|---|
| For use indoors | Flame-proof, heat-resistant, strong |
| For use underground | Same as above |
| For use outdoors | Same as above, plus cold-resistant and sunlight-proof |

A separate ground wire **must** be provided for nonmetallic conduits.

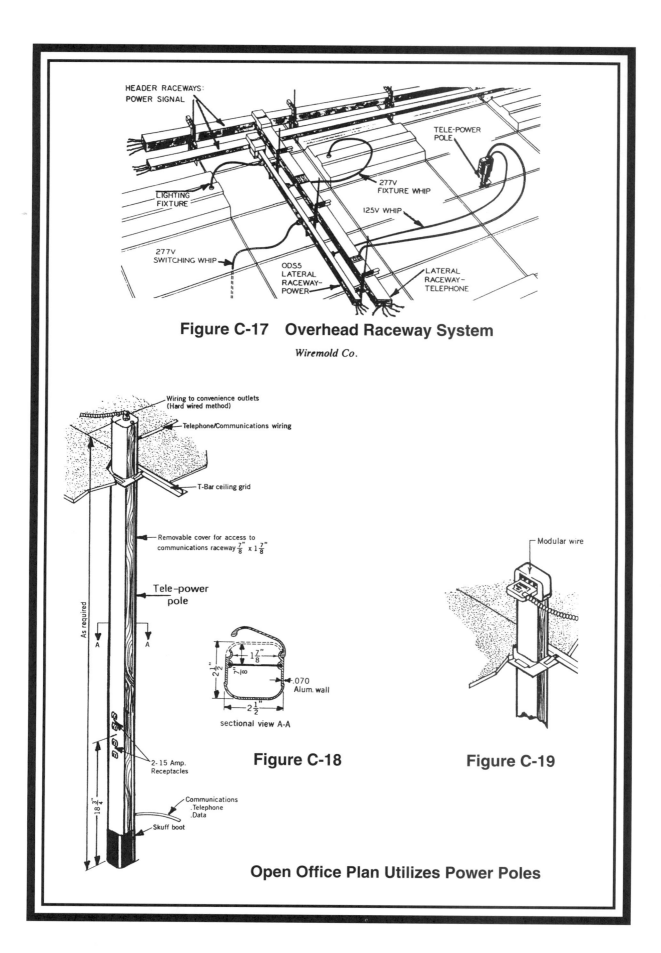

HEADER RACEWAYS:
POWER SIGNAL

TELE-POWER POLE

277V FIXTURE WHIP

125V WHIP

LIGHTING FIXTURE

277V SWITCHING WHIP

ODS5 LATERAL RACEWAY-POWER

LATERAL RACEWAY-TELEPHONE

**Figure C-17   Overhead Raceway System**

*Wiremold Co.*

Wiring to convenience outlets
(Hard wired method)

Telephone/Communications wiring

T-Bar ceiling grid

Removable cover for access to
communications raceway $\frac{7}{8}$" x $1\frac{7}{8}$"

Tele-power pole

As required

A     A

2-15 Amp. Receptacles

$18\frac{3}{4}$"

Communications
.Telephone
.Data

Skuff boot

$1\frac{7}{8}$"

$2\frac{1}{2}$"   $\frac{7}{8}$"

.070 Alum. wall

$2\frac{1}{2}$"

sectional view A-A

**Figure C-18**

Modular wire

**Figure C-19**

**Open Office Plan Utilizes Power Poles**

180

## C-26    WIREWAYS

Wireways are also referred to as:

*a.*    **Surface metal raceways**

*b.*    **Surface raceways**

*c.*    **Surface wireways**

*d.*    **Raceways, etc.**

They are metal channels with a removable cover which permits easy access to conductors and allows addition to or replacement of a wiring system (Figs. C-15 and C-16).

**They are used only for exposed applications and in dry locations.**

It is a good method of rewiring in existing buildings where new, concealed wiring systems may be difficult and more costly.

## C-27    CEILING RACEWAYS

This type of wiring system is less costly and more practical than floor raceways.

It consists of a network of steel or aluminum channels with a removable cover located above the tile ceiling (Fig. C-17).

The power and communication wiring systems are supplied to work areas below by movable floor-to-ceiling poles called **power poles** (Figs. C-18 and C-19).

The only disadvantage of this system is the unpleasant appearance of power poles all over the space.

## C-28    FLOOR RACEWAYS

There are two systems used for floor raceways:

1.    *Under-floor raceways.*    The distribution ducts are used under-floor to provide a wiring system for spaces above the floor (Fig. C-20).

2.    *Cellular floor raceways.*    A metal or precast concrete cellular flooring system is used for the floor structure. Each cell serves as a raceway carrying power and communication wiring services above the floor. The header ducts for feeding the wiring system in the cells run perpendicular to the cells (Fig. C-21).

**181**

Style SC
Bottomless
Trenchduct

Cast, One Piece
Preset Insert
(Triple Service)

Locking Flange

Self-Flashing Flange

Shear Stud

Communications
Or Data Cell

Power
Cell

Data Or
Communications
Cell

3" Walkerdeck
Raceway

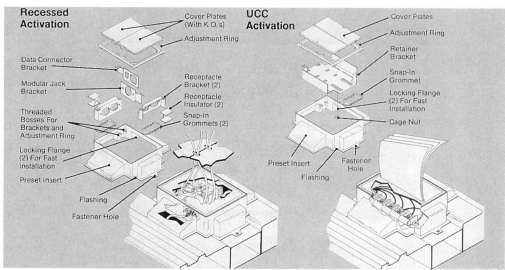

**Recessed Activation**

Cover Plates (With K.O.'s)

Adjustment Ring

Data Connector Bracket

Modular Jack Bracket

Receptacle Bracket (2)

Receptacle Insulator (2)

Snap-In Grommets (2)

Threaded Bosses For Brackets and Adjustment Ring

Locking Flange (2) For Fast Installation

Preset Insert

Flashing

Fastener Hole

**UCC Activation**

Cover Plates

Adjustment Ring

Retainer Bracket

Snap-In Grommet

Locking Flange (2) For Fast Installation

Cage Nut

Preset Insert

Fastener Hole

Flashing

## Figure C-20   Two-Level Underfloor Duct System

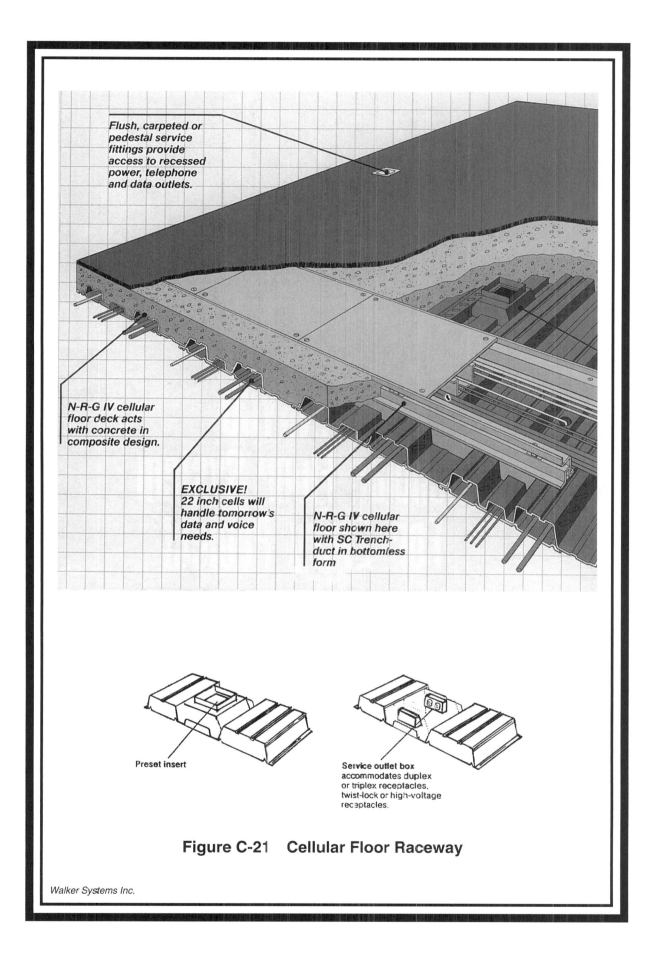

Flush, carpeted or pedestal service fittings provide access to recessed power, telephone and data outlets.

N-R-G IV cellular floor deck acts with concrete in composite design.

EXCLUSIVE!
22 inch cells will handle tomorrow's data and voice needs.

N-R-G IV cellular floor shown here with SC Trench-duct in bottomless form

Preset insert

Service outlet box accommodates duplex or triplex receptacles, twist-lock or high-voltage receptacles.

**Figure C-21    Cellular Floor Raceway**

# CONTENTS

# Part 5

# ELECTRIC SERVICE TO BUILDING(S)

# B

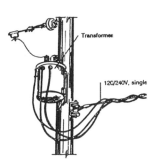

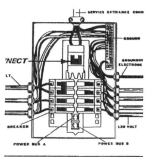

*A FLUSH-MOUNTED PANELBOARD*

## Electric Service to Building(s)

| CATEGORY | 1 | 2 | 3 | 4 | 5 |
|---|---|---|---|---|---|
| | | | Air Conditioning | | Ten-Year Percent |
| Type of Occupancy | Lighting | Misc. Power | Electric | Nonelectric | Load Growth |
| | | *Volt-amperes per sq. ft.* | | | |
| Auditorium | | | | | |
|    General | 1.0–2.0 | 0 | 12–20 | 5–8 | 15–35 |
|    Stage | 20–40 | 0.5 | | | |
| Art gallery | 2.0–4.0 | 0.5 | 5–7 | 2.0–3.2 | 15–35 |
| Bank | 1.5–2.5 | | 5–7 | 2.0–3.2 | 20–40 |
| Cafeteria | 1.0–1.6 | 0.5 | 6–10 | 2.5–4.5 | 15–35 |
| Church & synagogue | 1.0–3.0 | 0.5 | 5–7 | 2.0–3.2 | 10–30 |
| Department store | | | | | |
|    Basement | 3–5 | 1.5 | | | |
|    Main floor | 2.0–3.5 | 1.5 | 5–7 | 2.0–3.2 | 35–60 |
|    Upper floor | 2.0–3.5 | 1.0 | | | |
| Dwelling 0–3000 ft² | 3.0 | 0.5 | | | 35–60 |
|    3000–120,000 | 0.4 | 0.15 | | | |
|    Above 120,000 | 1.5–2.5 | 2.0 | | | |
| Hospital | 1.0–3.5 | 1.5 | 5–7 | 2.0–3.2 | 35–60 |
| Hotel | 1.0–2.0 | 0.5 | | | |
|    Lobby | 1.0–1.5 | 1.0 | 5–8 | 2.0–3.5 | 15–50 |
|    Rooms | 2.0–3.0 | 5–20 | 3–5 | 1.5–2.5 | |
| Laboratories | 1.5–3.0 | 1.5 | 6–10 | 2.5–4.5 | 30–60 |
| Library | 1.0–2.0 | 0.5 | 5–7 | 2.2–3.2 | 15–30 |
| Medical center | 1.5–3.5 | 2.5 | 4–7 | 1.5–3.2 | 30–50 |
| Office building | 1.5–2.8 | 2.5 | 4–7 | 1.5–3.2 | 30–40 |
| Restaurant | | | 6–10 | 2.5–4.5 | 20–30 |
| School | 2.0–2.5 | 2.0 | 3.5–5.0 | 1.5–2.2 | 30–50 |
| Shops | 2.0–3.5 | 0.5 | | | |

**Figure B-1    Electrical Load Estimating**

# ELECTRIC SERVICE TO BUILDING(S)

## B-1 ELECTRICAL LOAD ESTIMATING

Before starting to design any structure(s), a preliminary electrical load calculation must be made to determine the approximate total kilowatts of electricity required for the building(s).

The utility company may not have adequate power available adjacent to the site of your project. The electric company preparation for providing power to your building(s) may consume many months. This is the reason for determining the power requirements at the start of the design stage.

## B-2 ELECTRICAL LOAD CALCULATION PROCEDURE

The following calculation procedure is for rough estimating purposes. The exact total electrical load required can be determined only after the electrical design is completed.

$$W = VI$$

**Step 1.** Determine the electrical power (**VI**) needed for your project by using Fig. B-1.

**Total gross area of building in sq. ft. × VI/sq. ft. given in Fig. B-1**

    *a.*  **Lighting:** Use category 1.

    *b.*  **Miscellaneous power:** Use category 2.

Miscellaneous power includes, but is not limited to the following items:

        1.  Convenience outlets

        2.  Data processing terminals

        3.  Portable heaters

        4.  Small motors, etc.

    *c.*  **Air conditioning, electric:** Use category 3.

        Includes all electrical loads necessary to properly operate all heating, ventilating, and air-conditioning (HVAC) systems.

    *d.*  **Air conditioning, nonelectric:** Use category 4.

        When a building requires only heating and ventilating (H.V.), use 65 percent of the value of the load given in category 4.

e.   ***Water and sanitary systems:*** Use category 4.

Includes all electrical loads needed for the following:

1.   Water pump

2.   Vacuum and air compressor pumps

3.   Ejectors and sump pumps

4.   Well pumps

5.   Water heaters and pneumatic tubes

6.   Fire pumps

In your building(s) you may have:

1.   All of the items listed above

2.   A few of the items listed above

As a rule of thumb, use 16.7 percent for each item listed supra.

f.   ***Vertical transportation systems***

Vertical transportation systems include the following:

1.   All types of elevators

2.   Moving stairs (escalators)

3.   Dumbwaiters

4.   Moving ramps

5.   Moving walks (horizontal transportation)

To obtain the electrical loads required to operate vertical transportation:

1.   See section on vertical transportation given in this book; design the elevator system for your building; then

2.   Contact a local representative of the elevator manufacturing companies. They will furnish you with information in regard to approximate power requirements without charge.

g.   ***Kitchen equipment***

1.   For residential building(s) and a small kitchen in any building(s) electrical loads can be estimated by using Fig. B-2. Use the column **"Volt-Amperes"** for typical connection values.

2. For a large kitchen, contact *only* the kitchen consultants. They will furnish you with the approximate estimate of the electrical loads necessary to operate equipment installed in the kitchen, as well as other information you may need. They normally don't charge you any fee for this service.

h. ***Special equipment:*** Special equipment includes but is not limited to:

1. Special equipment (large computers, printing, display area loads, etc.) used in any building(s)
2. Equipment used in hospitals and health centers
3. Equipment used in laboratories
4. Electrical loads required for workshops
5. Monument lighting
6. Parking and walks lighting
7. Electric ice-melting walks
8. Canopy electric heaters and lighting
9. Stage lighting, etc.

### HOW TO OBTAIN INFORMATION

For additional information in regard to items *f, g,* and *h* stated supra:

1. Call **"Sweet's Automated Buy Line"** (check the latest Sweet's catalogs for the 800 number) and ask for the telephone number of the local manufacturer's representative.
2. The manufacturer's representative will furnish you with the information you need without any charge.

**Step 2.** *a.* Add all required electrical loads **(R.L.).**

*b.* From Fig. B-1, category 5, use the largest number given under **"Ten-Year Percent Load Growth"** to obtain additional loads required for expansion **(E.L.).**

**Total electrical loads required = (R.L.) + (E.L.)**

**Step 3.** Contact the electric company serving your site of the project to obtain, *in writing,* the answer to which of the following applies:

- *a.* Power supply from them will be adequate for your structure.
- *b.* Power is not available for your project. In this case they will furnish you with the following:
    1. The length of time needed to provide you with the required power
    2. The cost for bringing power to the site of your building(s)

**B-3    ELECTRICAL LOAD CALCULATION EXAMPLE**

*Problem*

Calculate the electrical load for a church.

Total gross area is 8650 sq. ft.

A small kitchen will contain built-in oven, range top, dishwasher, and refrigerator.

The church will have central electric air conditioning.

*Solution*

**Step 1.** *Lighting.*   Using Fig. B-1 (category 1) 1 to 3 W/sq. ft. and using 2 W per sq. ft.

8650 sq. ft. × 2 W/sq. ft. = 17,300 W or **17.3 kW**

**Step 2.** *Miscellaneous power.*   Using Fig. B-1 (category 2) 0.5 W/sq. ft.

8650 sq. ft. × 0.5 W/sq. ft. = 4325 W or **4.325 kW**

**Step 3.** *Air conditioning* (HVAC).   Using Fig. B-1 (category 3) 5 to 7 W/sq. ft. and using 6 W per sq. ft.

8650 sq. ft. × 6 W/sq. ft. = 51,900 W or **51.9 kW**

**Step 4.** *Kitchen.*   Using Fig. B-2:

| | |
|---|---|
| Built-in oven | 4,500 W |
| Range top | 6,000 W |
| Dishwasher | 1,200 W |
| Refrigerator | 300 W |
| Total | 12,000 W or **12 kW** |

**Step 5.** Add steps 1, 2, 3, and 4:

$$17.3 \text{ kW} + 4.325 \text{ kW} + 51.9 \text{ kW} + 12 \text{ kW} = 85.525 \text{ kW}$$

**Step 6.** Find the 10-year percent load growth using Fig. B-1 (category 5) which indicates 10 to 30 percent.

Using 30 percent:

$$85\,525 \text{ kW} \times .3 = 25.66 \text{ kW}$$

**Step 7.** Add steps 5 and 6:

$$85.525 \text{ kW} + 25.66 \text{ kW} = 111.185 \text{ kW} \approx 120 \text{ kW}$$

**The approximate total electric load required is 120 kW.**


**B-4    ELECTRIC POWER LINES**

Electric power lines in the avenues, streets, roads, etc., are divided into two categories:

1.    *Overhead* **(OH).**    Power lines are located on the top of electric poles.

2.    *Underground* **(UN).**    Power lines are placed underground.

Electric service to a building(s) may be:

1.    *Overhead service drop.*    The service lines can be extended to your building(s) according to your design.
    a.    Overhead (OH) (Fig. B-3).
    b.    Underground (UN) (Fig. B-4)—This service costs approximately 20 to 50 percent more than overhead, but it is more attractive, reliable, and has a longer life.

2.    *Underground service lateral.*    The service lines are extended to the building(s) underground from underground power lines (Fig. B-4).

For residences or small buildings, the electric company will install the transformer before the electric service enters the property (Fig. B-3) for overhead or (Fig. S-11) for underground service.

For medium and large buildings a transformer is located within the property before the meter.

| Appliance | Volt-Amperes | Volts | Wires | Circuit Breaker |
|---|---|---|---|---|
| Range | 12,000 | 115/230 | 3 #6 | 60 A |
| Oven (built-in) | 4,500 | 115/230 | 3 #10 | 30 A |
| Range top | 6,000 | 115/230 | 3 #10 | 30 A |
| Dishwasher | 1,200 | 115 | 2 #12 | 20 A |
| Waste disposer | 300 | 115 | 2 #12 | 20 A |
| Refrigerator | 300 | 115 | 2 #12 | 20 A |
| Washing machine | 1,200 | 115 | 2 #12 | 20 A |
| Dryer | 5,000 | 115/230 | 3 #10 | 30 A |
| Portable heater | 1,300 | 115 | 2 #12 | 20 A |
| Television | 300 | 115 | 2 #12 | 20 A |
| Fixed lighting | 1,200 | 115 | 2 #12 | 20 A |
| Air conditioner ¾ hp | 1,200 | 115 | 2 #12 | 20 A or 30 A |

**Figure B-2   Residential Electrical Equipment**

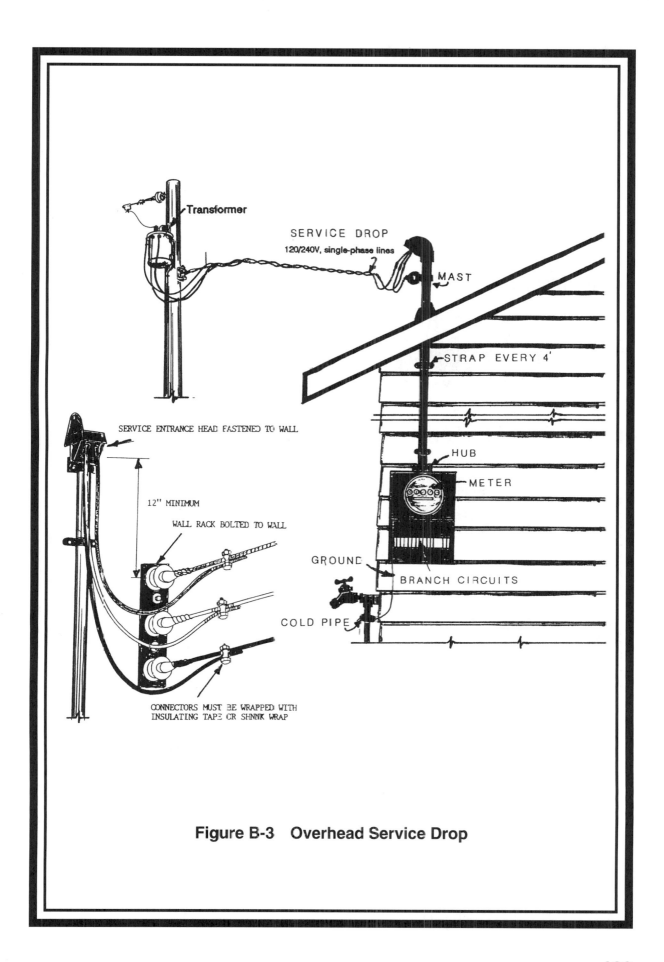

**Figure B-3  Overhead Service Drop**

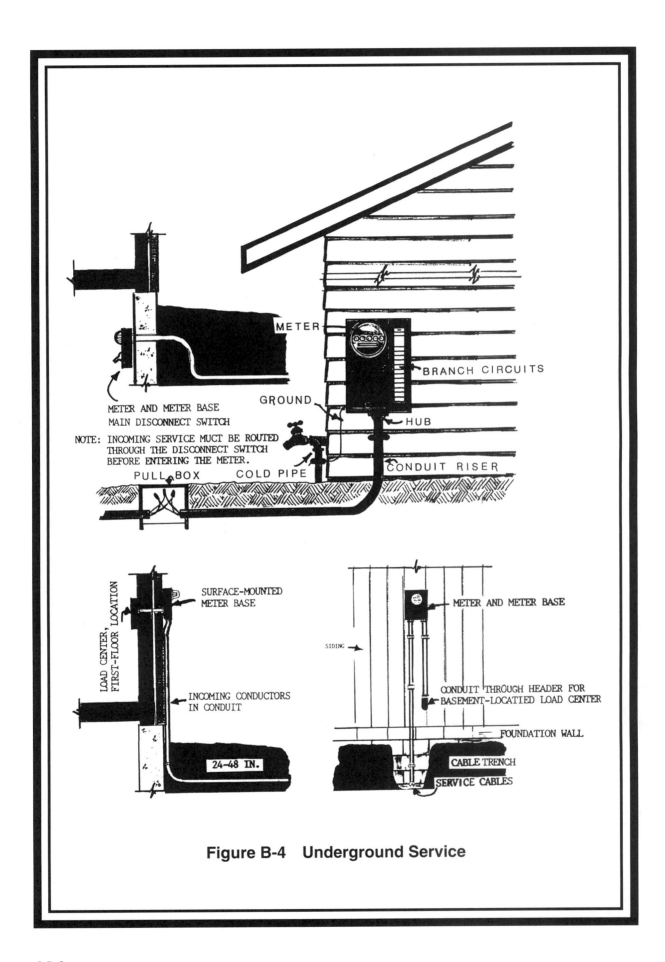

METER

BRANCH CIRCUITS

GROUND

HUB

METER AND METER BASE
MAIN DISCONNECT SWITCH

NOTE: INCOMING SERVICE MUCT BE ROUTED
THROUGH THE DISCONNECT SWITCH
BEFORE ENTERING THE METER.

CONDUIT RISER

PULL BOX    COLD PIPE

LOAD CENTER, FIRST-FLOOR LOCATION

SURFACE-MOUNTED
METER BASE

INCOMING CONDUCTORS
IN CONDUIT

24-48 IN.

METER AND METER BASE

SIDING

CONDUIT THROUGH HEADER FOR
BASEMENT-LOCATIED LOAD CENTER

FOUNDATION WALL

CABLE TRENCH
SERVICE CABLES

**Figure B-4   Underground Service**

194

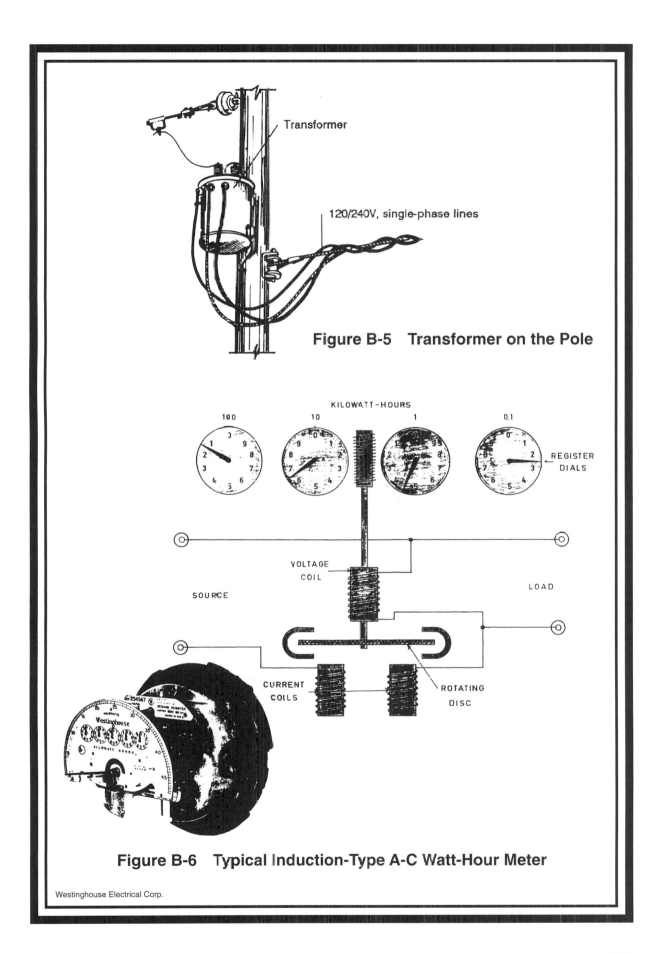

Transformer

120/240V, single-phase lines

**Figure B-5   Transformer on the Pole**

KILOWATT-HOURS

100          10          1          0.1

REGISTER
DIALS

SOURCE                    LOAD

VOLTAGE
COIL

CURRENT
COILS

ROTATING
DISC

**Figure B-6   Typical Induction-Type A-C Watt-Hour Meter**

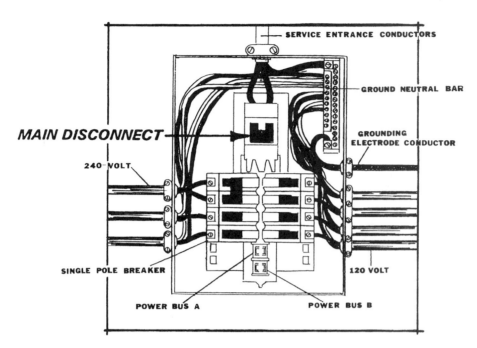

Figure B-7    A Flush-Mounted Panelboard

Figure B-8    Heavy-Duty Fused Three-Pole Three-Wire Switch

196

## B-5 ELECTRIC METER

The electric meters are installed and controlled by the utility company in order to measure the amount of electric power used in a building(s).

The electric meter is installed as follows:

*a.* **Outside on the exterior wall of building(s)**

*b.* **Inside the building accessible to a meter reader**

A meter is located between the service line and the main switch or panelboard with main breakers.

The typical induction-type **AC** kilowatt-hour meter consists of two types of coils (Fig. B-6):

1. *Current coils.* Located below the rotating disc.
2. *Voltage coils.* Located above the rotating disc.

The coils are connected to the electric power entering from the service lines and the electric load being used in the circuit branches within and outside the building(s).

The induction created when electric power is used causes the disc to rotate.

The speed of the rotating disc is proportional to the power being used.

**The number of rotations is counted on the dials, which are calibrated in kilowatt-hours.**

## B-6 MAIN SWITCH

It is also known as *service switch, service disconnect, main disconnect,* and *main breaker.*

The purpose of the main switch is to bridge, break, or disconnect the flow of all electricity to the building(s), except emergency equipment in the event of fire and emergency or for major repairs of the electrical system.

A main switch is installed in two different ways:

1. **Main switch is located within the panelboard** (Fig. B-7). This type is used in residential and small buildings.
2. **Main switch is installed as a separate device.** This type is used in medium and large buildings (Fig. B-8).

**197**

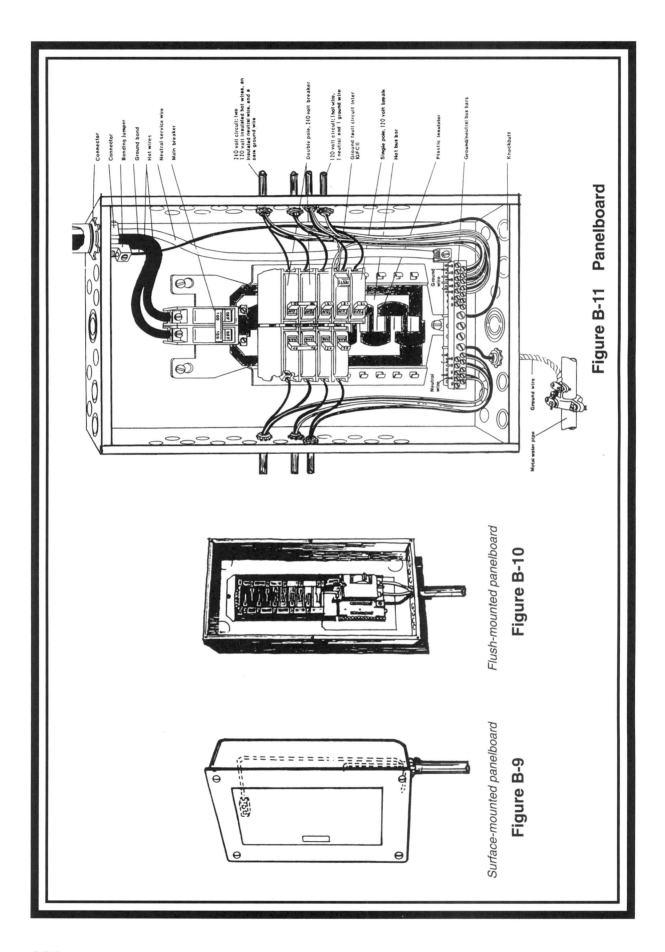

Connector
Connector
Bonding jumper
Ground bond
Hot wires
Neutral service wire
Main breaker
240 volt circuit: two 110 volt insulated hot wires, an insulated neutral wire, and a bare ground wire
Double pole, 240 volt breaker
120 volt circuit: 1 hot wire, 1 neutral and 1 ground wire
Ground fault circuit Inter IGFCI1
Single pole, 120 volt breake
Hot bus bar
Plastic insulator
Ground/neutral bus bars
Knockouts

Ground wire

Neutral wire

Metal water pipe    Ground wire

**Figure B-11    Panelboard**

*Flush-mounted panelboard*  **Figure B-10**

*Surface-mounted panelboard*  **Figure B-9**

198

A main switch is always located between a meter and circuit breaker cabinet.

A main switch is rated by current and voltage; also by the numbers of poles and throw.

## B-7   PANELBOARDS

These are also known as *electric panels* and *load centers.*

Panelboards consist of a metal cabinet containing overcurrent protection and other devices.

A panelboard is a distribution center for receiving electrical power and feeding the branch circuits.

Panelboards are classified into two mounting types:

1.   *Flush mounting.*   It is recessed flush with the finish-wall surface (Fig. B-9).
2.   *Surface mounting.*   It is surface-mounted usually on solid walls or columns (Fig. B-10).

Panelboards are also classified in two categories with regard to overcurrent devices:

1.   *Circuit breaker*
2.   *Fused*

Panelboards contain a main breaker (main switch) and main fuses to receive circuit-protective devices such as circuit breakers or fuses which serve the branch circuits containing electrical devices (lighting, outlets, appliances and motors, etc.).

A panelboard containing a circuit breaker is shown in Fig. B-11.

## B-8   CIRCUIT BREAKER

A circuit breaker is the most widely used device for overcurrent protection.

All conductors in a circuit branch are designed to carry a specific amount of current (15, 20, 25, 30 A, etc.), which is called the *ampacity* of the conductor. When the ampacity of the conductor exceeds the design ampacity because of

1.   **Overloading the circuit**
2.   **Problem in the circuit**—conductors' contact with each other (short circuit, unintentional grounding), etc.

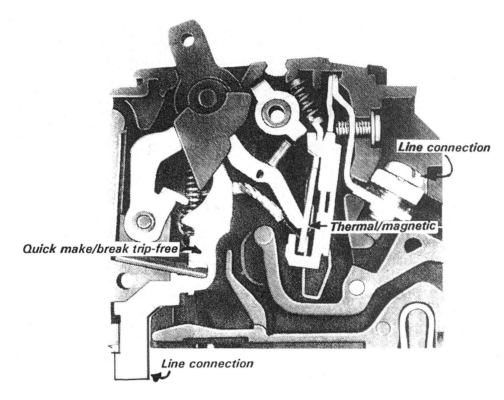

**Figure B-12   Molded-Case Circuit Breaker**

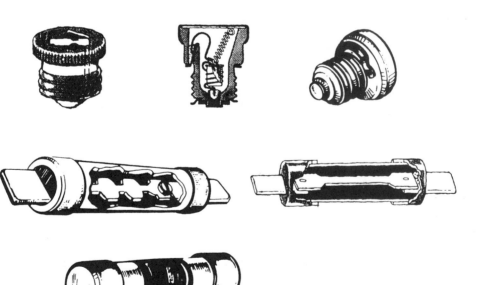

**Figure B-13   Types of Fuses**

the circuit breaker opens automatically (trips) and stops the flow of current in the branch circuit.

After the problem is corrected, the circuit breaker switch is turned on (Fig. B-12).

## B-9    FUSES

Before circuit breakers were introduced to the market, fuses were used exclusively as over-current protection. They are the simplest circuit-protective devices and are used in some buildings today.

Fuses are made in a variety of shapes and ratings.

They have two terminals which are connected together with a fusible metal (Fig. B-13).

Fuse metal is designed for a specific ampere (15, 20, 25 A, etc.). When the ampacity of the conductors feeding the circuit branch exceeds the design ampacity, the fusible metal melts and stops the current from flowing in the circuit.

A fuse is a one-time device and has to be replaced with a new one each time the fusible metal melts.

## B-10    GROUNDING AND GROUND FAULT PROTECTION

Grounding and ground fault protection in all wiring systems are used to safeguard from electrical shock the people who are using electrically operated devices and equipment.

## B-11    GROUNDING

Grounding connects the electrical system's neutral wire to the moist or wet ground.

There are several ways to accomplish this goal:

1.    **Connect the ground wire to cold-water piping in the building.**
2.    **Connect to a ground rod or ground plate placed underground which is moist or wet.**

In all buildings grounding is provided at the panel ground bus (Fig. B-14).

**NEC** and all electrical codes require that the electric service to the building *must* be grounded. They also require that all lighting, outlets, appliances, devices, and equipment, etc., ***must*** be properly grounded to provide ground path (Fig. B-14).

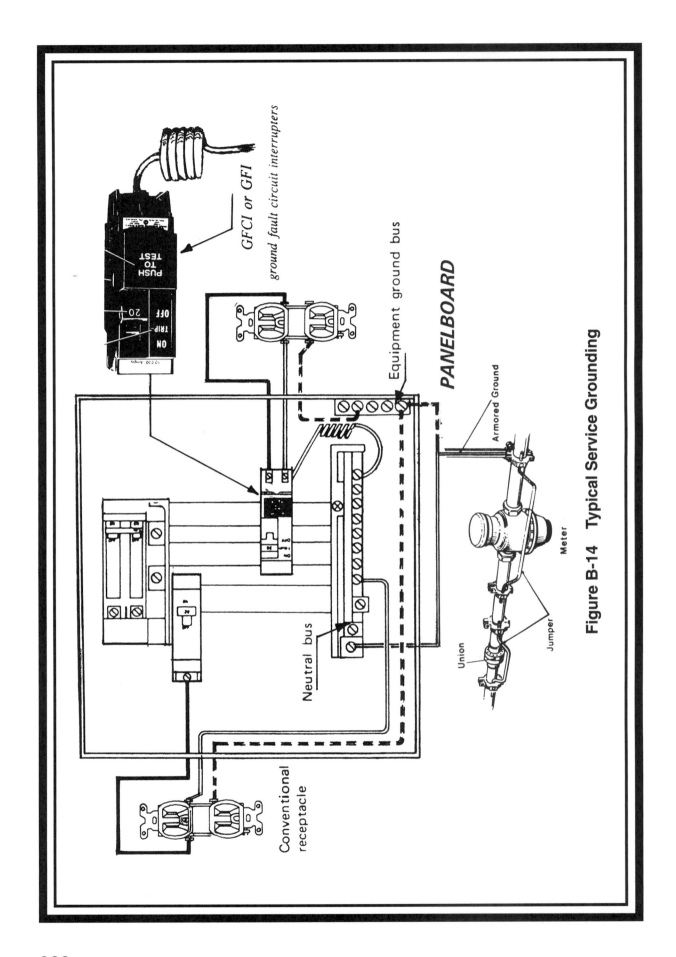

GFCI or GFI
ground fault circuit interrupters

PUSH TO TEST

TRIP

ON    OFF    20

10 000 Amps

Equipment ground bus

*PANELBOARD*

Neutral bus

Conventional receptacle

Armored Ground

Meter

Union

Jumper

**Figure B-14   Typical Service Grounding**

Manufacturers of appliances recommend that appliance housing be grounded to a cold-water pipe, and they supply appliances with a three-wire plug. Fig. B-15 gives an example of this type of grounding connection.

## B-12    GROUND FAULT PROTECTION

A person standing on dry wood, rubber, or other insulating material may feel the electric shock from a defective electrical device, but since that person has no direct contact with the ground, the shock may not be severe.

However, if the floor is wet (as in a kitchen, bathroom, outdoors, etc.) the electric shock can be severe or fatal.

**The resistance of the human body is approximately 1000 ohms** (some studies indicate 1000 to 4000 ohms).

For 120-V alternating current, 60 Hz, the person's resistance is as follows:

$$I = \frac{V}{R}$$

$$I = \frac{120\ V}{1000\ ohms} = 0.12 \text{ amp or } 129 \text{ ma}$$

**ma = milliamperes**

The overcurrent requirement to open a 15-A circuit breaker is approximately 148 ma.

The effect of alternating current (60 Hz) in a person is as follows:

| | |
|---|---|
| **15 to 30 ma** | **May freeze a person to a circuit** |
| **50 to 100 ma** | **May cause ventricular fibrillation** |
| **100 to 200 ma** | **Definite ventricular fibrillation** |
| **over 200 ma** | **Can cause contraction of muscle and may stop the heart** |

**The amount of 149 ma needed to open one 15-A circuit breaker has already caused the injury to the person.**

This is the reason for using ***ground fault circuit interrupter*** (GFCI) (Fig. B-16) in all locations which are or have a tendency to become moist and wet (kitchens, bathrooms, moist basements, outdoors, etc.) (Fig. B-16).

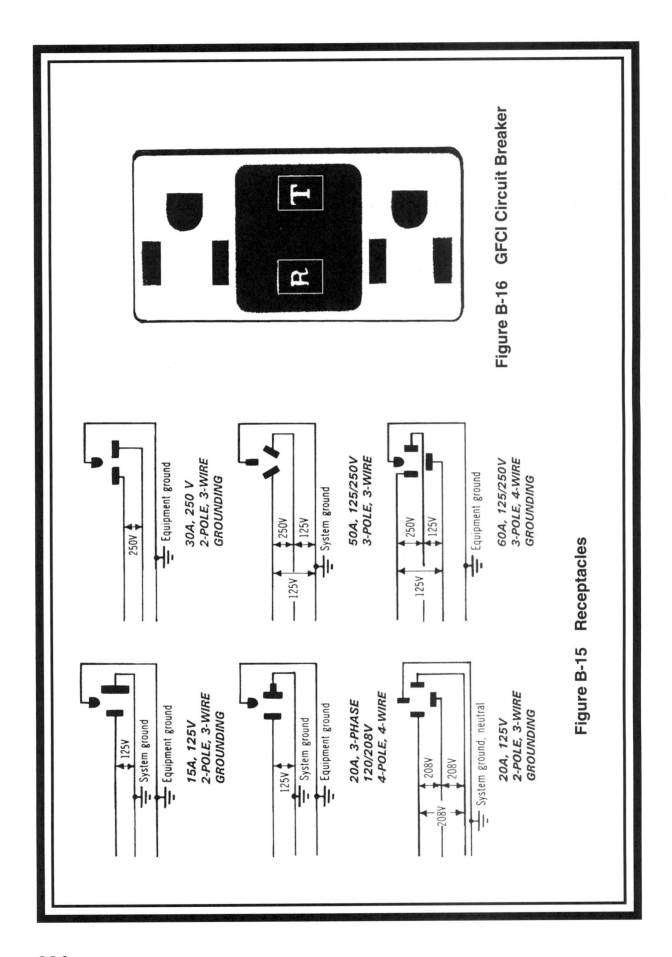

**Figure B-16  GFCI Circuit Breaker**

30A, 250 V
2-POLE, 3-WIRE
GROUNDING

50A, 125/250V
3-POLE, 3-WIRE

60A, 125/250V
3-POLE, 4-WIRE
GROUNDING

15A, 125V
2-POLE, 3-WIRE
GROUNDING

20A, 3-PHASE
120/208V
4-POLE, 4-WIRE

20A, 125V
2-POLE, 3-WIRE
GROUNDING

**Figure B-15  Receptacles**

In **NEC** and many local codes, the use of **GFCI** in moist or wet areas is mandatory.

It is also a good practice to use *ground fault interrupter* **(GFI)** for all appliances and equipment.

### B-13    GFCI AND GFI—HOW THEY OPERATE

**GFCI** and **GFI** operate on the same principle as any other receptacle—with one exception: at the same time it is supplying power, it monitors the current.

**Incoming current = outgoing current**

If outgoing current is less than incoming current, a current leakage is at work. This current leakage is called *ground fault.*

When this happens within one-half of a second, the **GFCI** or **GFI** will open the circuit and eliminate the shock hazard.

### B-14    SWITCHGEAR AND SWITCHBOARDS

A switchboard is an insulated metal cabinet 1½ to 4½ ft. deep with a width and height dependent on the number of circuits. It contains the following:

1.    **Primary disconnect switches**
2.    **Secondary feeder switches**
3.    **Overcurrent devices, circuit breakers, or fuses**

*Switchgear* is a term used to describe the switchboard when it is used for power which carries an above-600-V service.

Switchgears are located in a room (normally located in the basement) called the *electrical switchgear room.*

A proper natural ventilation system of about **1 sq. in. per kVA** of capacity should be provided only for this system.

# CONTENTS

## Part 6

# ELECTRICAL WIRING DESIGN

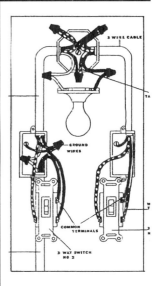

## Electrical Wiring Design

## Branch Circuit Design

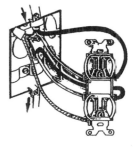

# ELECTRICAL WIRING DESIGN

**W-1 WIRING SYSTEMS**

Before starting to design a wiring system, review the local electric code or **NEC** for their requirements pertaining to your building. In the country or region that may have no code requirements, use **NEC** as your guide.

**A wiring system starts with the utility company's underground or overhead electric power lines which supply high-voltage and low-current power to:**

1. A required-size **step-down** *transformer.*
2. From the transformer feeder, deliver the necessary power to a **building's** *meter.*
3. From the meter, the feeders enter into the **building's** *main switch.*
4. The feeders from the main switch enter into the *panelboard.*
5. The power in the panelboard, after passing through the overcurrent protections for each individual circuit, enters into the *branch circuits.*
6. **The branch circuits deliver 15, 20, 30, 40, and 60 A to power the lights, outlets, appliances, and equipment within the building.**

**Wiring systems used in buildings may be classified in two groups:**

1. Residential and small buildings' (up to three stories) wiring systems
2. Medium-size and large buildings' wiring systems

**In residential or small buildings the** *panelboard* (B-7) is the central distribution of all electrical power used in the building.

All *circuits* start at and return to the panelboard with two or three wires.

In this case, only 1 meter is required; if 2 or more meters are used. after each meter a panelboard will be installed, either next to each other, or they can be installed in the area they are serving.

**In medium-size and large buildings the panelboard** may serve several floors or one floor (or portion of a floor), depending on the building, load, and code requirements.

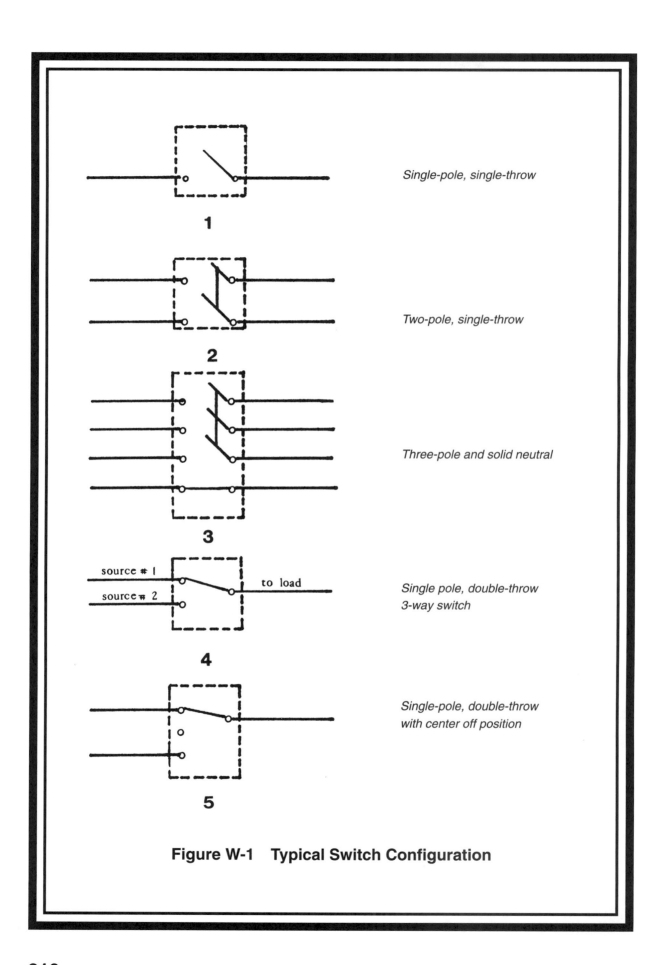

1 — *Single-pole, single-throw*

2 — *Two-pole, single-throw*

3 — *Three-pole and solid neutral*

4 — source # 1 / source # 2 / to load — *Single pole, double-throw 3-way switch*

5 — *Single-pole, double-throw with center off position*

**Figure W-1   Typical Switch Configuration**

## W-2    DEVICES USED IN WIRING SYSTEMS

In principle, the devices used in the wiring systems can be classified in two groups:

1.   *Switches*

2.   *Outlets*

Switches are devices for opening and closing or for changing the connection of a circuit.

Outlets are used in wiring systems from which electric current is diverted to supply an electric load to devices, appliances, and equipment.


## W-3    SWITCHES CLASSIFICATION

Switches are classified in accordance with the work they perform in wiring systems, as follows:

*a.*   ***Number of poles** (P) **and number of throws** (Fig. W-1).*   They are available in 1, 2, 3, 4, and 5P and single-throw, double-throw, or triple-throw.

*b.*   ***Current rating.***   The amount of current a switch can carry continuously (15, 20, 30 A at 120 V, or 120/277 V, etc.).

*c.*   ***Voltage class.***   The switches are rated for 250 V, 600 V, or 5 kV.

*d.*   ***Horsepower (HP).***   Switches used for motors are rated by HP.

*e.*   ***Heavy duty (HD).***   Switches used for constant interruption.

*f.*   ***Fusible.***   The switch contains a fuse(s).

*g.*   ***Enclosure.***   All separately enclosed switches are installed in metal cabinets.


## W-4    SWITCHES' OPERATING HANDLES

Operating handles of switches are called:

*a.*   ***Toggle types.***   Commonly used, and they create a small noise.

*b.*   ***AC quiet and mercury types.***   They are noiseless.

*c.*   ***Key, rocker, rotary, push, torch, and tap-plate types.***


## W-5    TYPES OF SWITCHES

There are many types of switches used in buildings. Some of the commonly switches used are as follows:

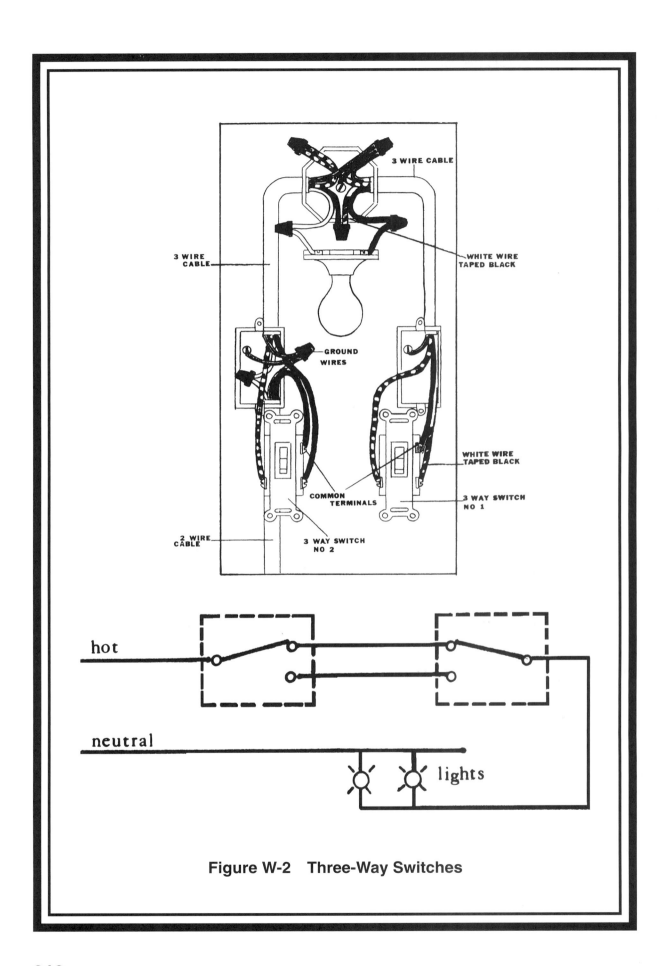

**Figure W-2   Three-Way Switches**

3 WIRE CABLE

WHITE WIRE TAPED BLACK

3 WIRE CABLE

GROUND WIRES

WHITE WIRE TAPED BLACK

COMMON TERMINALS

3 WAY SWITCH NO 1

2 WIRE CABLE

3 WAY SWITCH NO 2

hot

neutral

lights

1.  **Toggle switch** for 15, 20, and 30 A.

2.  **Contact switch** for 15 and 20 A. They are double-throw center off.

3.  **Top-plate switch** for 15 and 20 A.

4.  **Dimmer switch** for 5-A incandescent lamps.

5.  **Press switch** for 15 and 20 A. They have a lighted button in the off position.

6.  **Timer switch** 20 A. They are spring-wound for up to 60 minutes.

7.  **Key switch** for 15 and 20 A. Operates with a key.

8.  **Tumbler-lock-controlled switch.** These are used for high-security areas.

9.  **Programmable switch** 15 A. These are used to control the circuit by command.

### W-6 LOW-VOLTAGE SWITCHING

Also known as **low-voltage control switch** and **remote control switching.**

This type of switching permits flexibility for the control of lighting in a building, especially for a residential building.

The voltage requirement for this system is 24 V.

Low-voltage, low-current wires are connected between the light switch and the relay center.

When the light switch is turned on, it sends a signal with low voltage to the relay center.

This signal activates the relay switch, which opens the line voltage of 120 V to the light and allows the light to go on.

The chief advantages of this system are:

*a.* **Saving in wiring costs**

*b.* **Eliminates possible electric shock at the switch**

### W-7 THREE-WAY SWITCHES

A three-way switch is commonly used for:

*a.* A room with two entrances

*b.* Long hallways

*c.* Central hallways with stairs going up or down

*d.* Garages with entrances to the building and to outside, etc.

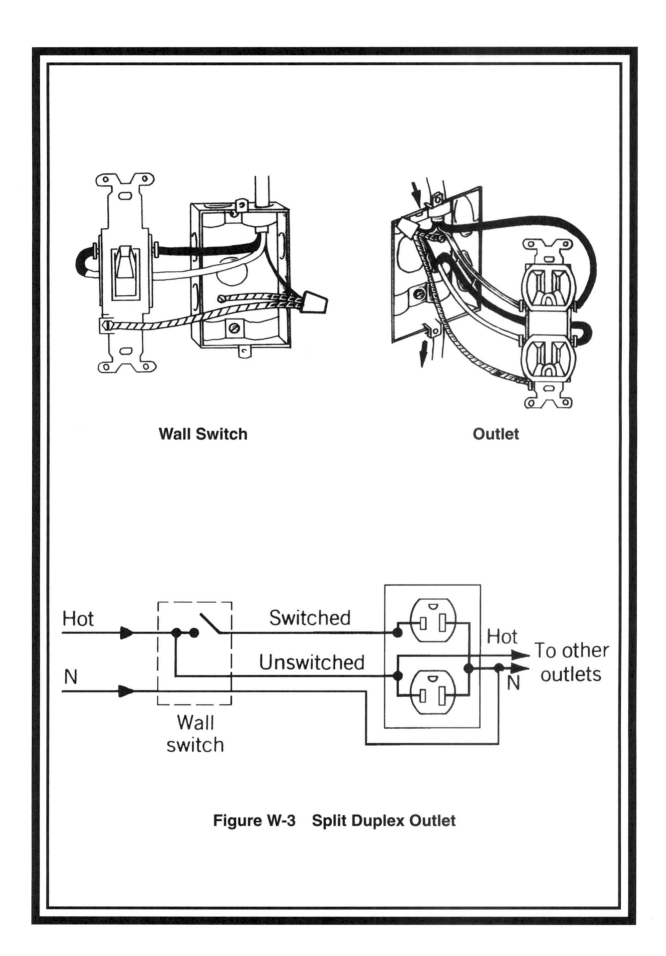

**Wall Switch**

**Outlet**

Hot

Switched

N

Unswitched

Hot

To other outlets

N

Wall
switch

**Figure W-3   Split Duplex Outlet**

In the above cases, it may be necessary to provide two switches to operate one or more lights, and to be able to turn on the light(s) on one side and turn them off on the other side. To accomplish this goal, a three-way switch is used.

Three-way switches use two single-pole double-throw switches, as shown in a wiring diagram (Fig. W-2).

## W-8 FOUR-WAY SWITCHES

Four-way switches are used for special areas which definitely must have three switches in three different locations to operate one or more lights. Their use is limited because of extensive wiring requirements.

## W-9 OUTLETS

Outlets are also called *receptacles, duplex receptacles, duplex outlets, convenience outlets,* and *receptacle outlets.* They may be classified as follows:

a. **Duplex outlets.** Two outlets in one box. These are commonly used.

b. **Single outlets.** They are generally used for appliances and equipment to prevent additional plug-ins.

c. **Ground fault interrupter (GFI) outlets.** Used for moist and wet areas (please see B-11, B-12, and B-13).

d. **Waterproof outlets.** Duplex outlets placed in a waterproof box generally used outside the building.

e. **Floor outlets.** Used in the floor with a protective cover.

f. **Safety outlets.** They are generally covered with a plastic cap when they are not in use.

g. **Switch-controlled outlets.** They are used in rooms (bedrooms, living rooms, etc.) where table lamps are used, and they are operated by a switch(es).

h. **Split-duplex outlets.** These are the same as the switch-controlled outlets, except that only one outlet is controlled by a switch (Fig. W-3).

| Appliance | Volt-Amperes | Volts | Wires | Circuit Breaker |
|---|---|---|---|---|
| Range | 12,000 | 115/230 | 3 #6 | 60 A |
| Oven (built-in) | 4,500 | 115/230 | 3 #10 | 30 A |
| Range top | 6,000 | 115/230 | 3 #10 | 30 A |
| Dishwasher | 1,200 | 115 | 2 #12 | 20 A |
| Waste disposer | 300 | 115 | 2 #12 | 20 A |
| Refrigerator | 300 | 115 | 2 #12 | 20 A |
| Washing machine | 1,200 | 115 | 2 #12 | 20 A |
| Dryer | 5,000 | 115/230 | 3 #10 | 30 A |
| Portable heater | 1,300 | 115 | 2 #12 | 20 A |
| Television | 300 | 115 | 2 #12 | 20 A |
| Fixed lighting | 1,200 | 115 | 2 #12 | 20 A |
| Air conditioner ¾ hp | 1,200 | 115 | 2 #12 | 20 A or 30 A |

**Figure W-4   Residential Electrical Equipment**

# BRANCH CIRCUIT DESIGN

**W-10    BRANCH CIRCUIT DESIGN**

A branch circuit may be classified into two categories:

1.    *SPECIAL-USE CIRCUITS*

2.    *GENERAL-PURPOSE CIRCUITS*


**W-11    SPECIAL-USE CIRCUITS**

They are also known as *controlled circuits.*

A special-use circuit branch is provided to accommodate only one appliance or equipment which consumes over 10 A (a range, oven, dryer, boiler, central HVAC, or air conditioner over ¾ HP, etc.). A list of some appliances and equipment is given in Fig. W-4.

All appliances and equipment must be properly grounded per manufacturer's instructions and electrical code requirements.


**W-12    GENERAL-PURPOSE CIRCUITS**

Branch circuits used for residential and small buildings serve both lighting fixtures and out-lets, and separate branch circuits are used for appliances and equipment.

Since the drawings for branch circuits are not complicated, they are usually incorporated into the floor plan(s) of the building.


**W-13    DESIGN CRITERIA FOR CIRCUIT BRANCHES**

Circuit branches can be classified into two divisions:

1.    **Circuit branch for residential and small buildings up to three stories**

2.    **For medium and large buildings (please see W-21)**


**W-14    CIRCUIT BRANCH FOR RESIDENTIAL**

**A. BRANCH CIRCUITS**

1.    Limit the general circuit branch load to 15 and 20 A. It is a good practice to **use 15 A** *only* **in residences.**

**NEC** particularly states that total **loads on branch circuits are not to exceed 80 percent of circuit rating.**

**NEC** does not specify the number of outlets used in a branch circuit in residential applications. However, it is a good practice to allow 240 watts maximum for each duplex outlet.

Therefore:

**Number of duplex outlets used on a 15-A** branch circuit at 120 V (80 percent of circuit rating used):

> **15 A $\times$ 0.8 = 12 A**
>
> **12 A $\times$ 120 V = 1440 W**
>
> **1440 W $\div$ 240 W/outlet = 6 outlets**

**Number of duplex outlets used on a 20-A** branch circuit at 120 V:

> **20 A $\times$ 0.8 = 16 A**
>
> **16 A $\times$ 120 V = 1920 W**
>
> **1920 W $\div$ 240 W/outlet = 8 outlets**

2. **Use six duplex outlets in a 15-A circuit branch, and eight duplex outlets in a 20-A circuit branch.**

3. **Do not use the branch circuit exclusively for lighting.**

4. **It is a good practice to use two branch circuits to serve each room, basement, playroom, and garage.**

## B. LIGHTING

1. For lighting in a residence, the **NEC** requires **3 W/ft$^2$ up to 3000 ft$^2$** and **0.4 W above 3000 ft$^2$.**

In living rooms and bedrooms lights are usually plugged into duplex receptacles. Three-way light bulbs require 150 W; therefore, the power in a duplex outlet is sufficient.

In dining rooms, kitchens, playrooms, hallways, and any space which requires a fixed lighting fixture, **calculate 3 W/ft$^2$.**

*Example*

a.   What is the total power required for lighting in a 12- by 14-ft dining room?

b.   How is this power calculated in a 15-A circuit branch?


*Solution*

a.   12 ft × 14 ft × 3 W/ft$^2$ = 432 W for lighting

b.   A 15-A branch circuit has a six-duplex outlet at 240 W each. Substitute a two-duplex outlet for lighting.

Therefore, the branch circuit will contain lighting fixture(s) plus four outlets.


2.   Provide at least one light in each closet with a switch outside the door.

3.   Provide one split-duplex outlet controlled by a switch in the proper places in rooms which have no overhead lights, such as bedrooms and living rooms.


## C. DUPLEX OUTLETS

1.   Duplex outlets should be installed *max.* **12 ft o.c.**

2.   **Any wall 2 ft or more** should have duplex receptacles.

3.   In a bathroom, at least on one wall, **a GFI duplex outlet** should be installed.

4.   At least one **waterproof outlet** should be installed outside of the building.

5.   A minimum of one **switch-controlled outlet** should be provided in a basement, utility room, attic, and in a craw space.

6.   One outlet should be provided in a hallway. If the hallway is more than 15 ft long, install one outlet for each 15 ft of length.

7.   Provide an individual circuit with the required power for each appliance and equipment using 10 A or more. Use a single outlet only to prevent additional plug-in. (Please see Fig. W-4 for information.)

8.   All outlets should be ground-type and must be grounded.

9.   Use ground fault interrupter **(GFI)** outlet in all locations which are or have a tendency to become moist or wet, such as bathrooms, kitchens, and basements.

UPPER LEVEL
MODEL - C

LOWER LEVEL
MODEL - C

### (Model C)  THE RIVIERA BI-LEVEL

4 nice sized bedrooms, 2 full baths, large living room, dining room, spacious family room with Fireplace* and sliding patio doors. Luxurious kitchen with dining area and General Electric Appliances*. Large open balcony, laundry center, storage and 1 (or 2)* car garage.

## Figure W-5   Design Circuit Branches for Example W-18

Architects & Engineers: Dadras International

## W-15 LOCATION OF OUTLETS

Before designing the locations of outlets on the walls, make a sketch of the possible location of furniture in order to **avoid placing the outlets under the bed, behind the couch, bookcases and cabinets, etc.** You may have to provide additional outlets, which is justified.

Outlets are placed commonly 15 inches above the finished floor, and 48 inches above the finished floor to accommodate the counters in kitchens, bathrooms, and workshops.

## W-16 LOCATION OF SWITCHES

Before locating the switches on the drawings, draw all door swings. **This will prevent locating switches behind the door when the door is opened.**

The height of wall switches is 50 inches from the finished floor. This is also within the height control for handicapped persons, which is 35 in. min., 54 in. max.

## W-17 LOAD CALCULATION

Add all the electrical loads used n the general circuit branch, and for special circuits add approximately 30 percent for future expansion. (Please see W-19.)

## W-18 EXAMPLE OF BRANCH CIRCUIT DESIGN

### Problem
Design circuit branches for the house shown in Fig. W-5 and show load calculations.

### Solution
The circuit branches designed for this house are shown in Fig. W-6:

1. Five general-purpose circuit branches are supplying power to lights and duplex outlets.
2. One general-purpose circuit branch is serving the unfinished lower level and the garage.
3. All duplex outlets were carefully placed in order not to be hidden under the bed or behind heavy furniture.
4. The bedrooms and living room are provided with a split-duplex outlet and a switch.
5. Door swings were drawn in order not to place the switches behind the doors.
6. All closets are provided with a light and a switch outside the closet.
7. Lights are provided in the front entrance, on the sun deck, and next to the garage entrance.

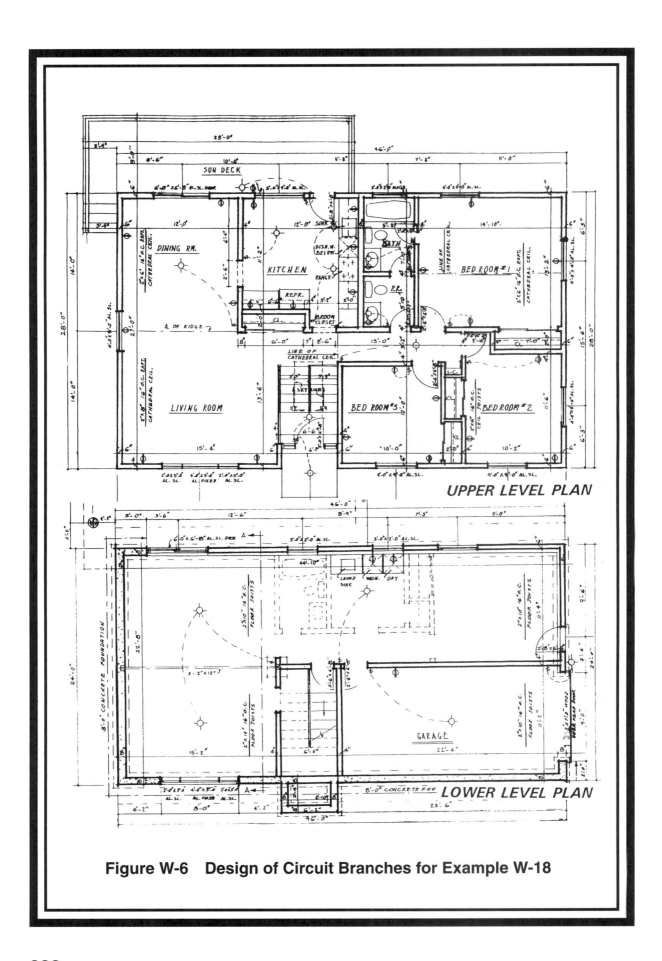

**Figure W-6    Design of Circuit Branches for Example W-18**

8. Exterior waterproof outlets are provided in the front, on the side, and on the sun deck.

9. Three-way switches are operating the lights in the main hall.

10. All duplex outlets in the kitchen, bathrooms, and laundry area are GFI-type and are so indicated.

11. Special-use circuits 7, 8, 9, 10, 11, and 12 are designated for the appliances and boiler by using Fig. W-4.

## W-19   ELECTRICAL LOAD CALCULATION FOR EXAMPLE W-18

| C. no. | Description | No. C.B. | V | VA load | Wire size AWG |
|--------|-------------|----------|---|---------|---------------|
| 1 | B.R. no. 1 & 2 | 1–15 A | 115 | 1725 | 2 no. 14 |
| 2 | B.R. no. 1, 2, 3, & hall | 1–15 A | 115 | 1725 | 2 no. 14 |
| 3 | B.R. no. 3, B. & K. | 1–15 A | 115 | 1725 | 2 no. 14 |
| 4 | K., D.R., hall, & sun deck | 1–15 A | 115 | 1725 | 2 no. 14 |
| 5 | L.R., D.R., hall, & outside | 1–15 A | 115 | 1725 | 2 no. 14 |
| 6 | Lower level & garage | 1–15 A | 115 | 1725 | 2 no. 14 |
| 7 | Refrigerator | 1–20 A | 115 | 300* | 2 no. 12 |
| 8 | Range | 2–30 A | 115/230 | 12000* | 3 no. 6 |
| 9 | Dishwasher | 1–20 A | 115 | 1200 | 2 no. 12 |
| 10 | Dryer | 2–15 A | 115/230 | 5000* | 3 no. 10 |
| 11 | Washing machine | 1–20 A | 115 | 1200 | 2 no. 12 |
| 12 | Boiler | 1–20 A | 115 | 1600* | 2 no. 12 |
| | Total | 14 | | 31,650 V.A. | |

C. no. = circuit number      No. C.B. = number and size of circuit breakers

V = volt      A = ampere      AWG = American Wire Gauge

* Check the actual VA rating for appliances and equipment you are using.

1. The numbers of circuit breakers needed is 14. Add 8 more for future expansion; therefore, the size of the *panelboard* must accommodate minimum 22 circuit breakers.

2. Load calculation:

    Load required          = 31,650 VA

    Add 30% for expansion

    $31,650 \times 0.3$         = 9.495 VA

    Total    41,145 VA

3. It is obvious that the total load will not be used at one time. Therefore, a *demand factor** should be applied **(NEC)**.

**Apply demand factor:**

**First 10,000 VA 100%** = 10,000 VA

**Above 10,000 VA 40%**

(41,145 VA − 10,000 VA) (0.4) = 12,458 VA

Total load      22,458 VA

22,458 VA ÷ 230 V = 97.64 A ≈ 100A

**The electrical service requirements for this house are 100 A at 115/230-V single-phase 3-wire.**

*** NEC** does not allow the demand factor for some equipment (compressor motors, fan-coil motors, etc.). Check the **NEC** requirements for more information.

*Note*

The wiring system used in this residence was nonmetallic sheathed **N.M.** cable (Romex).

It was cheaper than **AC** armored cable and less costly to install.

It was secured with the required spacing within 12 inches to 4½ feet.

It was installed carefully in order not to injure the cable.

This type of cable can be used in small buildings and residences, not to exceed three stories.

Check the local electric code before using this type of cable. Some local codes may not allow the use of **N.M.** cable.

**W-20 ELECTRICAL SYSTEMS FOR MEDIUM AND LARGE BUILDINGS**

For the design and preparation of construction documents in electrical systems for medium and large buildings you need the help of electrical engineer(s) "consultant(s)".

**Make sure:**

*a.* **They are capable of handling the work.**

*b.* **They are covered by adequate professional liability insurance.**

You are directly responsible to your client for the work they perform on your behalf.

The following guidelines are given for your information:

1. When the floor plans and sections are drawn with adequate dimensions (before detailing), make four mylar copies. Supply your consultants with three copies, and keep one copy in your file.

2. When any revision is made on the drawings, notify your consultants by telephone first and send a copy of the revision to them.

3. It is your responsibility to coordinate the electrical drawings with the reflected ceiling plans and the architectural, mechanical, and structural drawings.


**W-21 BRANCH CIRCUIT DESIGN FOR MEDIUM AND LARGE BUILDINGS**

In medium and large buildings, because of the extensive amount of lighting fixtures used, the general-purpose circuits are divided into two groupings:

1. Branch circuits to supply lighting fixtures only. They are shown in separate drawings which indicate the lighting fixture arrangements in each space, and the circuit which is supplying the power. This drawing(s) is generally called **lighting plan(s)**.

2. Circuits which are supplying the power for general-purpose outlets, and the circuits which are used for special purposes (such as appliances and equipment) are shown on separate drawing(s) commonly called **power plan(s)**.

All electrical system designs and calculations **must** be in accordance with local electrical codes or **NEC.**

The guidelines that follow are given for information only.

## W-22 ELECTRIC LOADS

1. Commercial buildings (such as a small office building or stores using a single-phase service to a large office building) demand a three-phase service.

2. All electrical loads used in commercial and industrial buildings are considered continuous unless specific facts and data are given to the contrary.

3. The main service may be a three-phase supply with a step-down transformer located in the building or outside on the premises.

4. The size of each panelboard is limited to a maximum of 42 overcurrent devices.

5. The rating of the main service is based on the total load required.


## W-23 DEMAND FACTORS

1. In many commercial and industrial buildings, no demand factors are allowed.

2. Demand factors are allowed for lighting loads in hotels, motels, warehouses, and hospitals.

3. In restaurants, the load for equipment may be subject to demand factors.


## W-24 GENERAL-PURPOSE OUTLETS

1. For office spaces up to 400 ft$^2$:

   a. **Supply one duplex outlet for every 10 linear ft of wall,** *or*

   b. **One duplex outlet for every 40 ft$^2$ (use *a* or *b* whichever is larger).**

2. For office spaces larger than 400 ft$^2$:

   a. **Calculate the total square feet of the area.**

   b. **Subtract 400 ft$^2$ from the total area.**

   c. **Use 10-duplex outlets for the first 400 ft$^2$ and divide the area by 110 ft$^2$ per outlet.**


*Example*

An office space is 30 ft wide and 90 ft long. How many duplex outlets are required?

*Solution*

30 ft × 90 ft = 2700 ft$^2$

2700 ft$^2$ − 400 ft$^2$ = 2300 ft$^2$

400 ft$^2$ ÷ 40 ft$^2$/out. = 10 out.

2300 ft$^2$ ÷ 110 ft$^2$/out. = 20.91 ≈ 21 out.

10 out. + 21 out. = 31 outlets are required

3. Limit the load of the branch circuit to 20 A and use only eight-duplex outlets per circuit. Each outlet is providing 240 W (please see W-14).

4. Provide 20 A at 120-V duplex outlet in the hallways approximately 45 feet o.c.

## W-25  SPECIAL-PURPOSE OUTLETS

For office and other buildings, draw the location of all the appliances, machines, and equipment on the drawings.

Obtain the power requirement for each equipment and provide adequate power.

Minimum-size copper conductors are as follows:

| 15 A | 20 A | 30 A | 40 A | 50 A |
|------|------|------|------|------|
| 14 AWG | 12 AWG | 10 AWG | 8 AWG | 6 AWG |

## W-26  LIGHTING

In order to calculate for the required power for lighting, we have to first design the lighting systems required.

**Please see Part 7, L-51 to L-75, "Lighting Design and Calculation."**

*a.* **First, design the lighting system for your building.**

*b.* **Then, calculate for the required *power.***

# CONTENTS

# Part 7

## LIGHTING SYSTEMS FOR BUILDING(S)

### Introduction to Lighting

### Lighting Systems for Building(s)

# Electric Light Sources

# Lighting Design and Calculation

## Local Lighting Calculation

## General Lighting Calculation

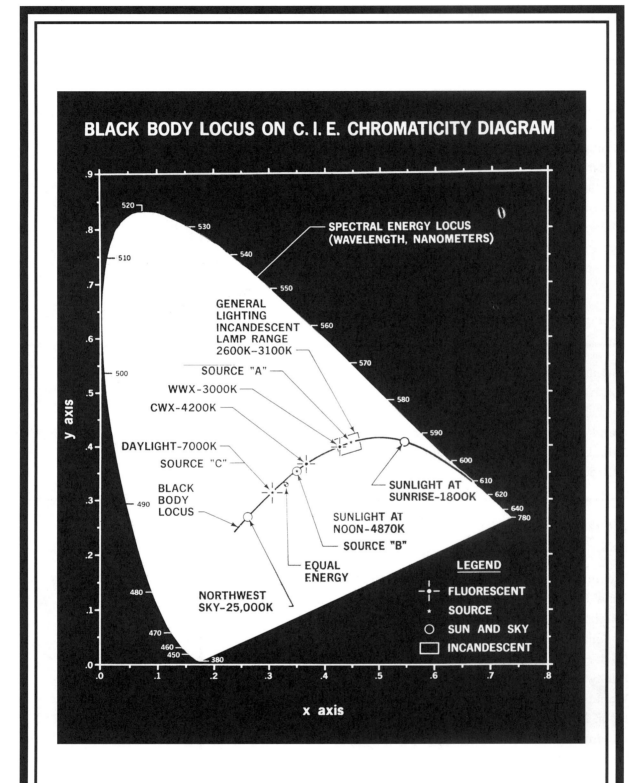

**Figure L-1**

# INTRODUCTION TO LIGHTING

## L-1   LIGHT AND ITS THEORY

Light is a form of energy, radiated as electromagnetic waves, which travels at a speed of 186,000 miles per second.

The theory of light was started by **C. Huygens** in 1690 and **Newton**'s corpuscular theory in 1704, which states:

> Light consists of minute particles emitted from luminous bodies; they travel through space at great speed and obey the laws of mechanics.

The theory of light was carried on by **T. Young** in 1801 and **J. Fresnel** in 1815.

**J. C. Maxwell** developed the electromagnetic wave theory in 1873, which states:

> Light is electromagnetic radiation covering an appropriate wavelength region.

The most famous experiment of **M. Morley** in 1881 led **Einstein** to develop the special theory of relativity in 1905. Einstein gave a formula for this effect, in which he presupposed:

> Light to be absorbed as discrete amount of energy called *quanta.*

**Modern quantum theory states:**

> Light is composed of photons, whose motions are guided by wavelike properties associated with them.

**Another version of quantum theory states:**

> The luminous bodies emit radiant energy in discrete bundles; these are ejected in straight lines and act upon the eyes to produce the sensation of light.

**Electromagnetic theory states:**

> The luminous bodies emit light in the form of radiant energy. This energy is transmitted in the form of electromagnetic waves.

None of the above theories have been conclusively proven by science.

**The Illuminating Engineering Society (IES)** in a simple and logical form defines light as:

*a.*   **Visually evaluated radiant energy**

*b.*   **A form of energy which permits us to see**

*c.*   **Radiant energy evaluated according to its capacity to produce visual sensation**

| | |
|---|---|
| **40,000 B.C.** | At the ending of the Paleolithic period, stone or shells of suitable shape were used for burning **animal oil for lighting.** |
| **4,000 B.C.** | Pottery lamps were used by Egyptians using **vegetable** or **animal oil and a wick.** |
| **600 B.C.** | **Torches** were supplemented in Greece with **pottery and metal lamps.** |
| | *Candlesticks* of the Hebrews were a support for a group of **float-wick lamps.** Its symbolical descendant is the eight-branched Hanukkah lamp. |
| | The primitive open-cruse types, using animal or vegetable oil, were commonly used from the Middle Ages up to the 19th century. |
| | **Candles** made of **beeswax** and **vegetable wax,** etc., were competing with lamps in Roman times and during the Middle Ages. |
| | The center of the **round wick** received very little oxygen from the air and, for this reason, these candles were smokey. |
| **Mid-18th century** | **Flat wicks** were introduced to the market. |
| | **Aime Argand** of France invented a **circular wick** with an open center and a **glass lamp chimney.** Whale oil was used in these lamps with no smoke. |
| **Late 18th century** | **Coal gas** was used for lighting. |
| **1802** | **Gas lighting** was used in London. |
| **1817** | Gas lighting was used in Baltimore. |
| **1823** | Gas lighting was used in New York State. |
| **1801** | Sir **Humphrey Davy** developed the **electric arc lamp.** |

| | |
|---|---|
| **1852** | Sir **George G. Stokes** discovered that fluorescence can be induced in certain substances by stimulation with ultraviolet light. |
| **1867** | **A. E. Becquerel** constructed a **fluorescent lamp** similar to those made today. |
| **1879** | **Charles F. Brush** produced **carbon-arc street lamps** which were used to illuminate the streets of Cleveland. Thereafter, they were used in other cities. |
| **1879** | **Thomas Alva Edison** developed the **incandescent lamp,** and it became widely used. |
| **1903** | **Peter Cooper Hewitt** invented the **mercury-vapor electric lamp.** |
| **1911** | **Georges Claude** invented the **neon lamp.** |
| **1913** | **Irving Langmuir** invented the **gas-filled incandescent lamp.** |
| **1920** | Fluorescent tubes were used for advertising signs. |
| **1938** | First practical **hot-cathode, low-voltage fluorescent lamp** was introduced to the market in the United States. |
| **1957** | **Tungsten-halogen** lamps, also called **quartz** lamps, were invented by General Electric Corp. |

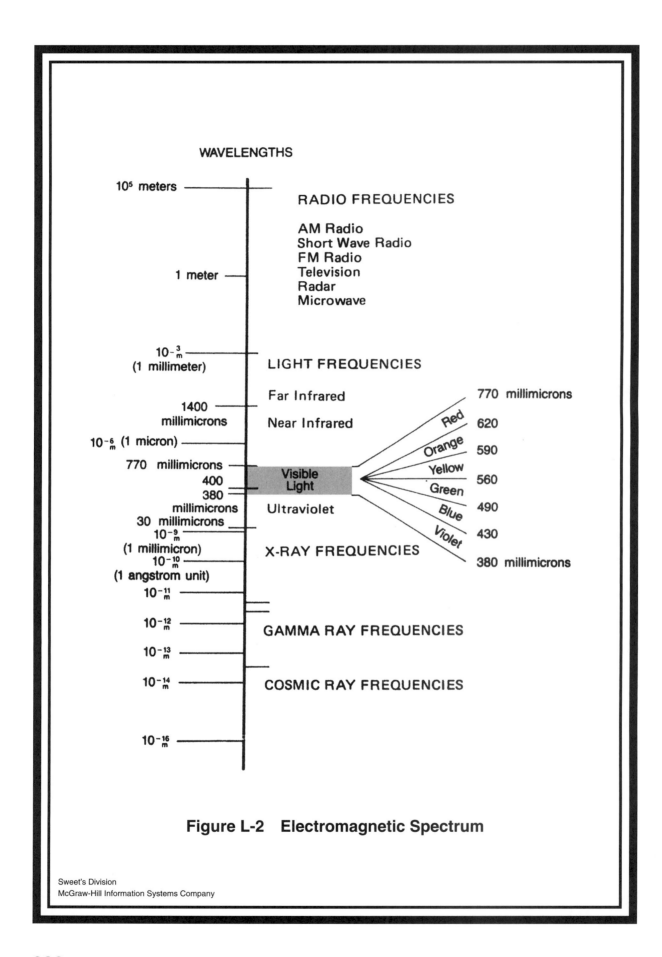

**Figure L-2   Electromagnetic Spectrum**

## L-3    VISIBLE AND INVISIBLE LIGHT

The effect of radiation of the wavelength of light in the human eye is limited between the ultraviolet and infrared sections of the spectrum of light.

| a. | *Invisible Light* | *Frequency** |
|---|---|---|
| | Cosmic rays | $10^{22}$ |
| | Gamma rays | $10^{20}$ |
| | X rays | $10^{18}$ |
| | Ultraviolet | $10^{16}$ |

| b. | *Visible light* [†] | *Wavelength*** |
|---|---|---|
| | Black light | 330 |
| | Violet | 380 |
| | Indigo | 400 |
| | Blue | 450 |
| | Green | 540 |
| | Yellow | 550 |
| | Orange | 640 |
| | Red | 700 |
| | Deep red | 780 |

| c. | *Invisible light* | *Frequency** |
|---|---|---|
| | Infrared | $10^{14}$ |
| | Radio or hertzian waves | $10^{10}$ |
| | Short wave | $10^{6}$ |

\* Frequency in cycles/second = Hertz (Hz)

\*\* Wavelength given is nanometers (metric system): nanometers (nm) = $10^{-9}$ meter.

[†] *Visible light* is the balance of several or all of the visible light (Fig. L-2).

## L-4    VISIBLE LIGHT

There are three types of visible light:

a.    **Natural phenomena light**

b.    **Temperature radiation light**

c.    **Luminescence light**

Visible light is observed and identified by the human eye as *white light.*

White light in reality is the combination of all colors of the visual spectrum such as violet, blue, green, yellow, orange, and red.

## L-5    NATURAL PHENOMENA LIGHT

Natural phenomena light is the radiation of the complete spectrum of light from:

a.    **Sun**

b.    **Stars**

c.    **Lightning**

d.    **Aurora borealis** and **aurora australis** are illuminous displays of various forms and colors seen in the night sky.

   The aurora borealis of the Northern Hemisphere is called **northern lights,** and is visible around the north regions.

   The aurora australis of the Southern Hemisphere is known as **southern lights,** and is visible around the south regions.

e.    **Bioluminence** is light produced from the oxidation of chemical compounds.

Other natural phenomena light is produced by the reflection of sunlight from:

a.   *Sky*

b.   *Moon*

c.   *Clouds*

d.   *Earth*

e.   *Bodies of water*

## L-6   TEMPERATURE RADIATION LIGHT

This is also known as *incandescence light.*

When a steel bar is heated, its free electrons become increasingly active.

**Up to 932°F only heat** radiation is released.

**Above 932°F infrared light** is produced.

**Above 1112°F visible light** with long wavelengths of red are released.

As the temperature increases, the shorter wavelengths appear, producing orange, yellow, and finally white.

Incandescence nucleus produces

**Heat + infrared light + visible light**

## L-7   LUMINESCENCE LIGHT

**Gas or solid\*** which is composed of a single valence atom (the outermost shell of an atom which has one to eight free electrons) raises its temperature when it receives electric current.

a.   **First, valence electron temporarily becomes energized.**

b.   **Then valence electrons return to their natural state and produce light.**

\* Cesium Cs,   fluorine F,   francium Fr,   hydrogen H,   lithium Li.

## L-8    ULTRAVIOLET (UV) LIGHT

Ultraviolet radiation is invisible light. It can be harmful or beneficial.

*a.*    **It can improve the intensity of some kinds of illumination.**

*b.*    **It may help the generation of vitamin D in the body.**

*c.*    **It is associated with the aging process of eyes and skin.**

*d.*    **It may embrittle textiles and paper.**


## L-9    INFRARED (IR) LIGHT

Infrared radiation is invisible light represented by its relation to heat.

*a.*    **It can be the cause of burns on skin and eyes.**

*b.*    **It is related to cataracts of the eye.**

*c.*    **It is used for heat and drying therapy.**

*d.*    **It can dehydrate fibers, etc.**


## L-10    COLORS OF LIGHT

The human eye is sensitive to the spectrum of visible light.

A small percentage of the total male population is faced with **color abnormalities** which are rare in the female population.

Color-vision defects can be inherited, for example, *color blindness*, or they can be caused by disease or poisoning called *acquired color defects*.

The lighting design requires the understanding of the relationships of primary and secondary colors.

1.    *Primary colors of light* (Fig. L-3)

    *a.*    **Red**

    *b.*    **Green**

    *c.*    **Blue**

Equal proportions of the three produce *white light.*

2. *Secondary colors of light* (Fig. L-4)

    *a.* **Magenta**

    *b.* **Cyan**

    *c.* **Yellow**

Equal proportions of the three or substraction of all colors produce *black.*

Other colors are produced by equal combinations of colors as follows:

    **Magenta** = blue + red

    **Cyan**　　= blue + green

    **Yellow**　= red + green

    **Red**　　 = yellow + magenta

    **Blue**　　= magenta + cyan

    **Green**　= yellow + cyan

The primary colors of pigment and dye absorb one of the primary colors of light and transmit or reflect the other two colors.

The perceived color of an object is the color of transmitted or reflected light.

Complementary colors are any two colors which, mixed in proper proportions, produce white:

    Blue + yellow　　= **white**

    Red + cyan　　　= **white**

    Green + magenta = **white**

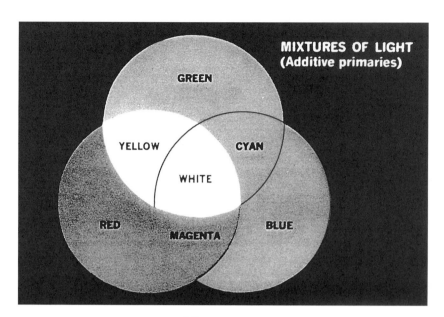

Figure L-3

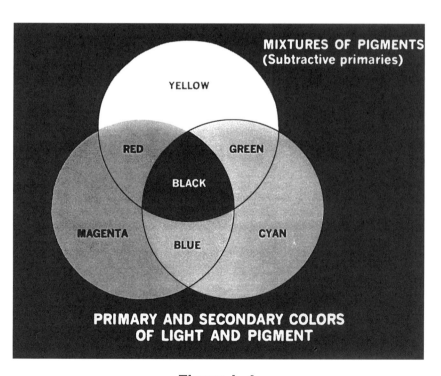

PRIMARY AND SECONDARY COLORS
OF LIGHT AND PIGMENT

Figure L-4

242

## L-11 COLOR TEMPERATURE

Color temperature may be described as the "***appearance of emitted light*** *as compared to* ***white light***."

Color temperature relates to the fact that when we heat a light-absorbing body it starts glowing from deep red to finally blue-white.

**Black ⟶ Deep red ⟶ Red ⟶ Orange ⟶ Blue-white**

The changes of color from black to blue-white are measured by a scale called ***color temperature***. **(Fig. L-5.)**

## L-12 KELVIN (K)

**W. T. Kelvin** (1824–1907), British mathematician and physicist, introduced the Kelvin scale (absolute scale) of temperature.

Kelvin color temperature relates to the physics of light, not the heat radiation.

The degree on the scale is called ***kelvin*** **(K).**

The color temperature assigned to artificial and natural light is given in Fig. L-5.

## L-13 RANGE OF COLOR TEMPERATURES

The color temperature starts with deep red light, gradually changing to orange → yellow → white → blue-white.

When the radiated light is balanced, such as direct sunlight, the light is at its whitest.

### LOW COLOR TEMPERATURES

Low color temperatures are related to ***yellowish light,*** which give the sense of warmth.

The lowest color temperature of light is the **yellow glow** of candlelight, which is **1800 K.**

### HIGH COLOR TEMPERATURES

The high color temperatures are related to ***bluish light,*** which gives the sense of coolness.

The highest color temperature of a light source is the sunlight of a clear blue sky which is over **10,000 K.**

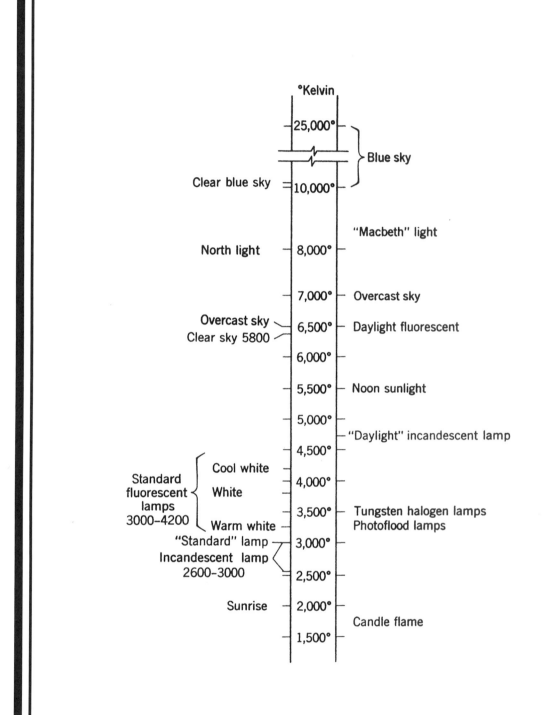

**Figure L-5   Color Temperature Scale**

244

# LIGHTING SYSTEMS FOR BUILDINGS

**L-14    ILLUMINATING ENGINEERING SOCIETY (IES)**

The Illuminating Engineering Society of North America was established in 1906 and continuously disseminates knowledge relating to the advancement of the art and science of illuminating engineering.

The **IES** has published many books and handbooks in the field of lighting to guide us in the proper design and application, and we should be thankful to them for their efforts in the last 88 years.

**L-15    ADEQUATE LIGHTING**

Our eyes demand proper and adequate lighting. When we read, write, draw, and work under *inadequate* amounts of light or *excessive* amounts of light, we feel tired or, after a short period of time become exhausted. Our efficiency will drop drastically, and we may experience headaches.

The physiological elements and components of the seeing process have a direct effect on *health* and *normal function.*

If we continuously work under improper lighting conditions for a long period of time, we may create irreparable problems in our sense of sight.

We architects have the moral obligation to design proper lighting for the people who will use the spaces within the buildings we have designed. Different rooms for different functions require a different and proper lighting level on the work plane. Remember:

**Inadequte lighting in a space**

is just as bad as

**excessive lighting in a space.**

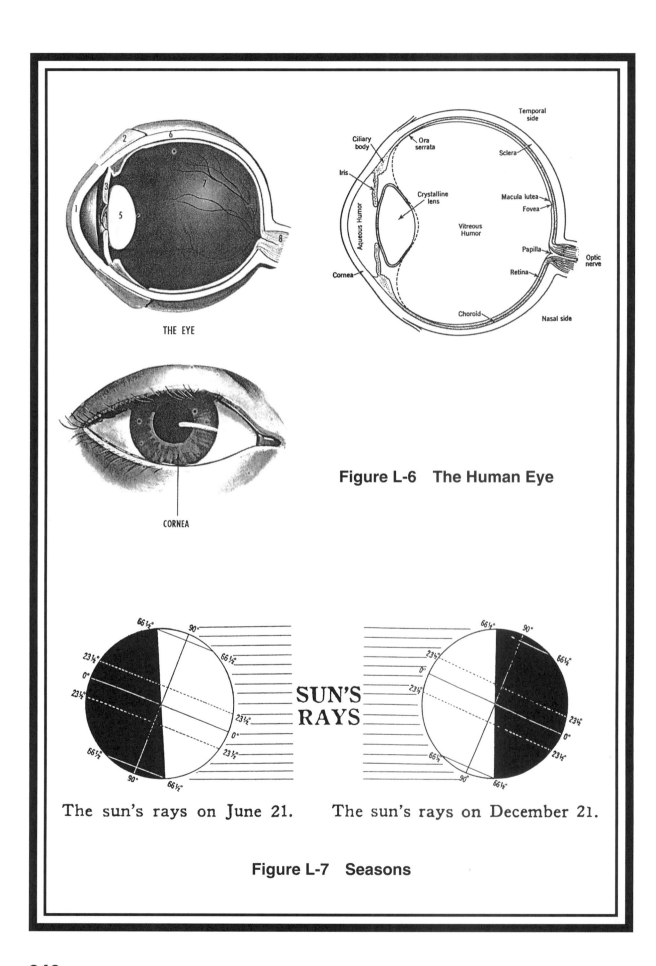

THE EYE

CORNEA

**Figure L-6   The Human Eye**

SUN'S
RAYS

The sun's rays on June 21.     The sun's rays on December 21.

**Figure L-7   Seasons**

## L-16    SENSE OF SIGHT

All the requirements for seeing are automatically controlled by the eye's organs (Fig. L-6).

The eye can quickly adapt to the changes in the lighting level.

All parts of the eye are working together in order to contribute an impulse to the brain, called *sense of sight.*

In the back of the crystalline lens a curvature plane is located, called the *retina.*

The retina is very sensitive to light, and is composed of *photoreceptors,* which are connected to the *brain* by the *optic nerve.*

The photoreceptors consist of:

a.    *Cones*

b.    *Rods*

The **cones** are located approximately at the center of the retina and transmit the images to the brain. The cones are not sensitive to low amounts of light, but they perceive color.

The **rods** are located on the outer portion of the retina. They are sensitive to light and do not respond to color.

The focusing of visual images on the **retina** is cone by the *crystalline lens,* which is controlled by the *ciliary muscle.* For near vision, the lens becomes round, and for distance vision the lens becomes flat by the tension produced in the ciliary muscle.

The amount of light entering the eye is controlled by the *iris* which has a center opening called the *pupil.*

The sensitivity of the retina controls the size of the opening of the **pupil.**

When the **lighting is inadequate,** the **pupil** tries to open as much as it can, and when the **lighting is excessive,** the **pupil** tries to close as much as it can. This creates a tension which may create discomfort and perhaps a headache.

Therefore, *inadequate lighting is just as bad as excessive lighting.*

*Example:* For reading and writing, 70 fc (footcandle) is required; less than 70 fc is inadequate, and more than 70 fc is excessive lighting.

We all know that a building must be designed to safeguard the *safety* and *health* of its occupants. It is a good practice to add *vision* to this phrase.

## L-17   LIGHTING

Lighting can be classified into two categories:

1.   *Natural daylighting*
2.   *Artificial lighting*

## L-18   NATURAL DAYLIGHTING

The natural light provided by the sun is bright and pleasant; however, it is not a stable light source, and it is constantly changing the space illumination requirements.

The factors affecting the sunlight illuminating the interior of buildings are (Fig. L-7):

a.   *Exterior effects*

Other buildings, terrain, and landscaping

b.   *Position of the sun*

In different seasons and latitudes

c.   *Clouds*

The sky covered by clouds

Staggered clouds which drastically change the intensity of the sun

In the United States above 40° latitude, southern orientation for windows will allow maximum sunlight in the winter and proper lighting.

Below 35° latitude, northern orientation for windows will allow proper lighting and prevent excessive solar heat from penetrating the interior of the building.

In some buildings (museums, art galleries, libraries, department stores, etc.) the effect of daylight may not be desirable. However, in other buildings the effect of daylight not only helps to reduce the cost and save electric energy, it also has a great physiological effect on the occupants.

The British **"Permanent Supplementary Artificial Lighting in Interior" (PSALI)** prescribes the use of daylight in buildings to supply a good portion of the building's lighting demands during the daylight hours.

Proper daylight illumination will not affect the visual performance.

Most European countries are taking advantage of daylight for illuminating their buildings.

We architects in the United States should follow their lead.

## L-19  ARTIFICIAL LIGHTING

From the Stone Age **Paleolithic period,** humans have been faced with artificial lighting problems and constantly have tried to discover a solution for it. Today we are faced with greater problems and tasks in artificial lighting.

The construction cost of lighting systems in a building is 3 to 5 percent of total construction cost.

All types of lighting fixtures and devices are at our disposal. We need to know

1. **Visional needs**
2. **Types of lighting**
3. **How they work**
4. **How to install them**
5. **How they are measured**
6. **How to design proper lighting**

in order to create a proper, economical, efficient, and pleasing visual environment within our building.

**Uniform Lighting**

**Figure L-8**

**Figure L-9   Special Effects**

## L-20    LIGHTING METHODS

Lighting methods can be divided into three types of applications:

1.    *General lighting*
2.    *Local lighting*
3.    *General and local lighting*

## L-21    GENERAL LIGHTING

General lighting is used in a general office, classrooms, lecture halls, meeting rooms, and any other similar space, which requires **uniform lighting** or a **blanket of light** on the horizontal work plane (top of desk, top of drafting table, etc.) throughout the space.

In general, common light fixtures containing fluorescent lamp(s) which are capable of producing uniform light are used (Fig. L-8).

To create special effects, incandescent lamps or high-intensity discharge lamps may be used (Fig. L-9). They are more expensive and have a shorter life than fluorescent lamps.

## L-22    LOCAL LIGHTING

Local lighting is also known as **supplementary lighting.**

It is used when a high intensity of illumination is required on special areas such as desks, drafting tables, walls, and displays.

For local lighting, incandescent lamps, fluorescent lamps, or high-intensity discharge lamps are used.

They can be installed (Fig. L-10) as follows:

1.    **Recessed in the ceiling**
2.    **Surface-mounted on the ceiling**
3.    **Hung from the ceiling**
4.    **As track lighting**
5.    **Placed on the desk, table, etc.**

**Figure L-10**

**Figure L-11   General and Local Lighting**

**L-23 GENERAL AND LOCAL LIGHTING**

General and local lighting is also called **combined general and local lighting.** This lighting method is used in the following circumstances:

1. When general lighting is designed for a specific visual task, but in the same space other special work or displays require higher intensity of illumination and need local lighting in addition to general lighting (Fig. L-11)

2. When an architectural design requires both general lighting and local lighting (Fig. L-12)

**L-24 TYPES OF LIGHTING SYSTEMS**

Lighting systems are classified into six types:

1. **Direct lighting**
2. **Indirect lighting**
3. **Semidirect lighting**
4. **Semi-indirect lighting**
5. **Direct-indirect lighting** or **General diffused lighting**

They may be used individually or in combinations of two or more, as required for visual tasks and architectural effects of the lighting system.

**L-25 DIRECT LIGHTING SYSTEMS**

In direct lighting systems distribution of luminance (Figs. L-13 and L-14) is as follows:

> **over 90 percent of the light is downward**
> **10 percent of the light is upward**

When distribution of luminance is from 0 to 90° it is called **spread lighting.**

In this type of lighting, the upper part of the walls and ceiling are illuminated by the light reflected from the floor, the lower parts of the walls, and the furniture.

If the higher part of the walls require higher intensity of illumination, wall washer reflectors or track lighting may be used.

This type of lighting is commonly used in general lighting design or concentrated local lighting.

**Figure L-12    General Diffuse Lighting or Direct-Indirect Lighting**

## L-26 INDIRECT LIGHTING SYSTEMS

In indirect lighting systems distribution of luminance (Fig. L-15) is as follows:

*over 90 percent of the light is upward*

*10 percent of the light is downward*

When distribution of illumination is from 90 to 130°, it is called ***spread lighting*** (Fig. L-15). It is used to illuminate the ceiling. **The ceiling height for this type of lighting is min. 9 ft 6 in.**

The angle of light may be narrowed in order to create concentrated illumination for special architectural effects.

The horizontal work plane of the room is illuminated by the reflection of light from the ceiling and walls.

a. **White and glossy ceilings cause the highest reflectance, about 80 percent.**

b. **Black and matte ceilings produce no reflectance.**

c. **The colors between black and white produce reflectance according to their brightness and finishes.**

This type of lighting normally produces uniform illumination with little or no glare.

For areas within the room requiring high-intensity illumination, local or supplementary lighting may be used.

## L-27 SEMIDIRECT LIGHTING SYSTEMS

In semidirect lighting systems distribution of luminance (Fig. L-16) is as follows:

*60 to 90 percent of the light is downward*

*10 to 40 percent of the light is upward*

This type of lighting produces a bright ceiling and walls and creates a pleasant atmosphere within the room.

It can create a great architectural effect, especially in the interior spaces without windows.

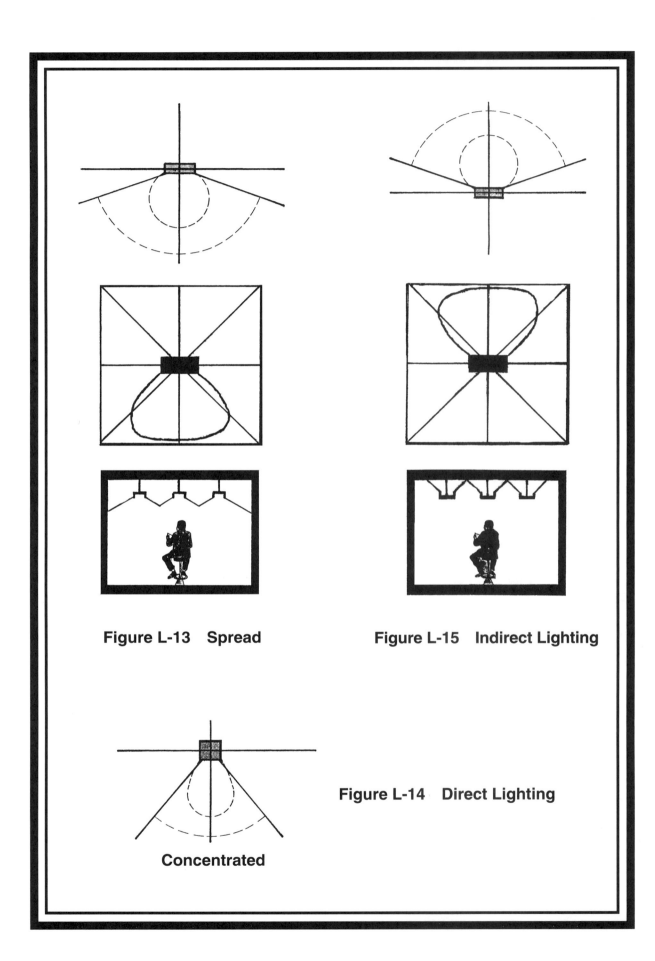

**Figure L-13   Spread**

**Figure L-15   Indirect Lighting**

**Concentrated**

**Figure L-14   Direct Lighting**

**L-28   SEMI-INDIRECT LIGHTING SYSTEMS (Fig. L-17)**

In semi-indirect lighting systems, distribution of luminance is exactly reverse of semidirect lighting:

> *10 to 40 percent of the light is downward*
>
> *60 to 90 percent of the light is upward*

It works on the same principle as semidirect lighting, except that the ceiling and upper walls of the room receive more illumination.

**L-29   DIRECT-INDIRECT LIGHTING SYSTEMS**

In direct-indirect lighting systems, distribution of luminance (Fig. L-18) is as follows:

> *50 percent of the light is downward*
>
> *50 percent of the light is upward*

This type of lighting reduces the contrast within a room and provides general and local illumination.

**GENERAL DIFFUSING LIGHTING SYSTEMS**

This type of lighting system is identical to direct-indirect lighting systems except that contrast and shadows created at the center for direct-indirect lighting are minimized to the outside.

They provide illumination in all directions and furnish uniform lighting for all functions. (Fig. L-19).

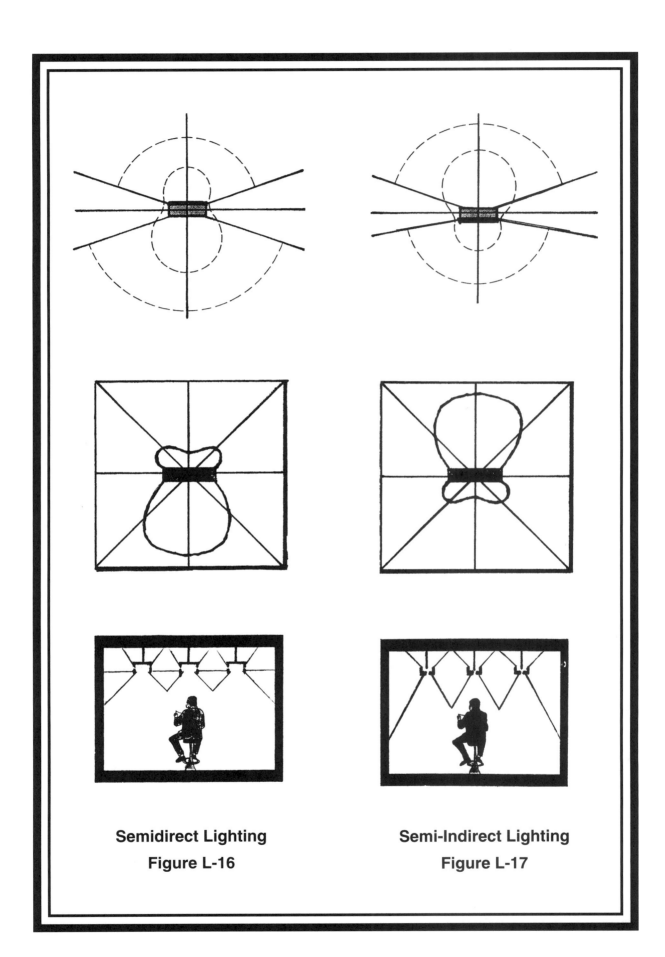

**Semidirect Lighting**

**Figure L-16**

**Semi-Indirect Lighting**

**Figure L-17**

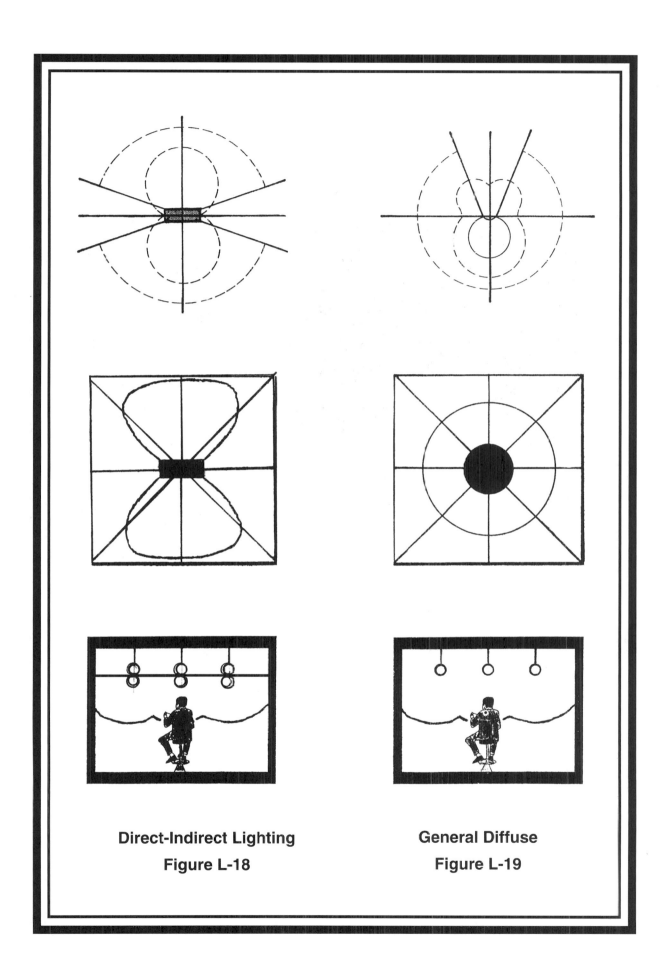

**Direct-Indirect Lighting**

**Figure L-18**

**General Diffuse**

**Figure L-19**

**Figure L-20   Incandescent Lamps**

# ELECTRIC LIGHT SOURCES

**L-30    ELECTRIC LAMPS**

The electric light sources (lamps) can be classified into six groups:

1.    **Incandescent Lamps** (L-32, L-33)

      **Tungsten-Halogen Lamps** (L-34)

2.    **Fluorescent Lamps** (L-35 to L-43)

      **Compact Fluorescent Lamps** (L-44)

3.    **High-Intensity Discharge (HID) Lamps** (L-45 to L-47)

4.    **High-Pressure Sodium (HPS-HID) Lamps** (L-48)

5.    **Low-Pressure Sodium (LPS) Lamps** (L-49)

6.    **Neon Lamps** (Fig. L-50)


**L-31    INCANDESCENT LAMPS**

The first practical incandescent lamp was invented by T. A. Edison in 1879.

The incandescent lamps (Fig. L-20) may be classified into two groups:

1.    *Incandescent filament lamps*

2.    *Tungsten-halogen lamps*


**L-32    INCANDESCENT FILAMENT LAMPS**

In an incandescent lamp the light is produced by heating the filament.

The hotter the filament, the more efficient it is in converting electricity to light.

When the filament operates hotter, its life becomes shorter, that is why the design of incandescent lamps is based on a balance between efficiency and life of the lamp.

This is the reason that lamps of equal wattage have different lumens and life rating.

They are very inefficient in converting electrical energy into visible light, approximately from **100 percent power received by a filament:**

|  |  |
|---|---|
| **72 percent goes to** | **Infrared heat** |
| **18 percent goes to** | **Heat** |
| **6 to 12 percent goes to** | **Light** |

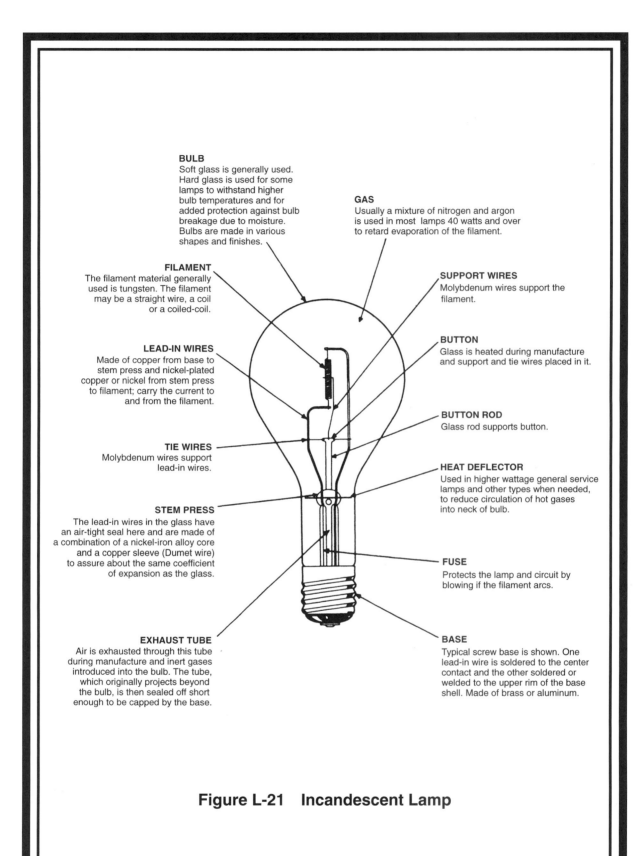

**BULB**
Soft glass is generally used. Hard glass is used for some lamps to withstand higher bulb temperatures and for added protection against bulb breakage due to moisture. Bulbs are made in various shapes and finishes.

**GAS**
Usually a mixture of nitrogen and argon is used in most lamps 40 watts and over to retard evaporation of the filament.

**FILAMENT**
The filament material generally used is tungsten. The filament may be a straight wire, a coil or a coiled-coil.

**SUPPORT WIRES**
Molybdenum wires support the filament.

**LEAD-IN WIRES**
Made of copper from base to stem press and nickel-plated copper or nickel from stem press to filament; carry the current to and from the filament.

**BUTTON**
Glass is heated during manufacture and support and tie wires placed in it.

**BUTTON ROD**
Glass rod supports button.

**TIE WIRES**
Molybdenum wires support lead-in wires.

**HEAT DEFLECTOR**
Used in higher wattage general service lamps and other types when needed, to reduce circulation of hot gases into neck of bulb.

**STEM PRESS**
The lead-in wires in the glass have an air-tight seal here and are made of a combination of a nickel-iron alloy core and a copper sleeve (Dumet wire) to assure about the same coefficient of expansion as the glass.

**FUSE**
Protects the lamp and circuit by blowing if the filament arcs.

**EXHAUST TUBE**
Air is exhausted through this tube during manufacture and inert gases introduced into the bulb. The tube, which originally projects beyond the bulb, is then sealed off short enough to be capped by the base.

**BASE**
Typical screw base is shown. One lead-in wire is soldered to the center contact and the other soldered or welded to the upper rim of the base shell. Made of brass or aluminum.

**Figure L-21   Incandescent Lamp**

The lamp temperature at peak varies by the wattage of the lamp: 100 to 500°F.

They commonly produce 2700 to 3200 K.

Illumination below 20 fc produces yellowish white light.

Incandescent filament lamps are available in many wattages, sizes, shapes, and colors.

The construction of a typical incandescent lamp is given in Fig. L-21.

The lamp bases are shown in Fig. L-22.

The lamp filament types are given in Fig. L-23.

## L-33    LONG-LIFE LAMPS

When the incandescent lamps are designed for higher voltage than they require to operate, their life span increases in direct proportion to the voltage difference.

The lamp rated at 130 V operating at 120 V will increase its life span by 300 percent; however, its lumen output is reduced to 78 percent.

### Example

We have an incandescent lamp which is designed to operate at 120 V. If the voltage applied is less than or more than 120 V, the following will result:

| Lamp operates at 115 V | Lamp operates at 125 V |
|---|---|
| (–5 V) | (+5 V) |
| Light decreases by 15% | Light increases by 16% |
| Power decreases by 7% | Power increases by 7% |
| Efficacy decreases by 8% | Efficacy increases by 8% |
| **Life increases by 72%** | **Life decreases by 42%** |

Light = Lumen (Lm)    Power = Watt (W)    Eficacy = Lumen/Watt (Lm/W)
Life = Hours of burning (h)

**Long-life lamps are expensive, and they are used in the areas which are hard to reach, such as high ceilings, etc.**

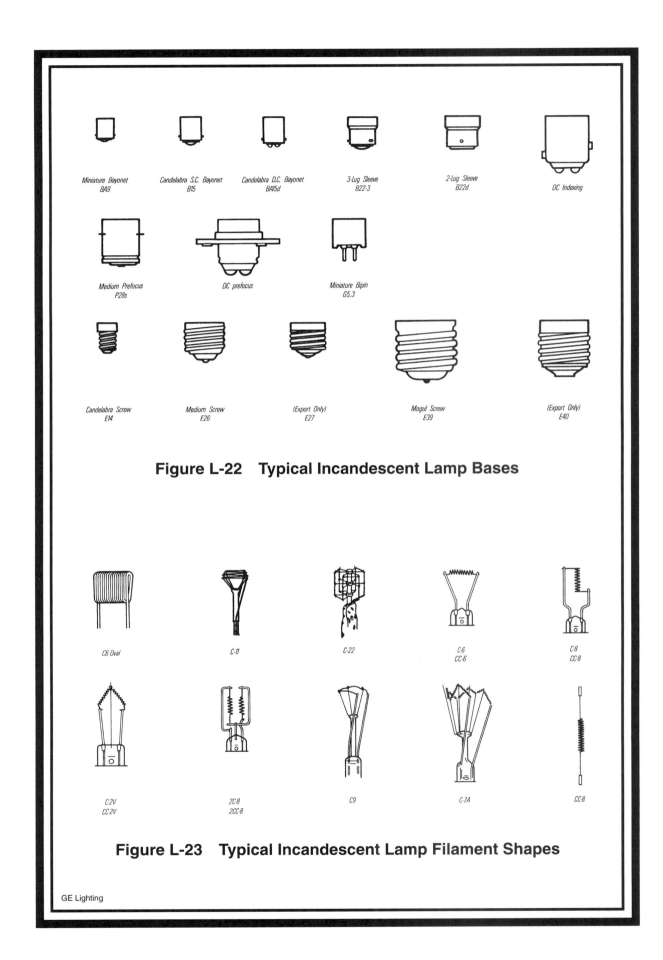

Miniature Bayonet
BA9

Candelabra S.C. Bayonet
B15

Candelabra D.C. Bayonet
BA15d

3-Lug Sleeve
B22-3

2-Lug Sleeve
B22d

DC Indexing

Medium Prefocus
P28s

DC prefocus

Miniature Bipin
G5.3

Candelabra Screw
E14

Medium Screw
E26

(Export Only)
E27

Mogul Screw
E39

(Export Only)
E40

**Figure L-22   Typical Incandescent Lamp Bases**

C6 Oval

C-11

C-22

C-6
CC-6

C-8
CC-8

C-2V
CC-2V

2C-8
2CC-8

C9

C-7A

CC-8

**Figure L-23   Typical Incandescent Lamp Filament Shapes**

**Figure L-24    Tungsten-Halogen Lamps**

# LAMPS: types, properties
## INCANDESCENT

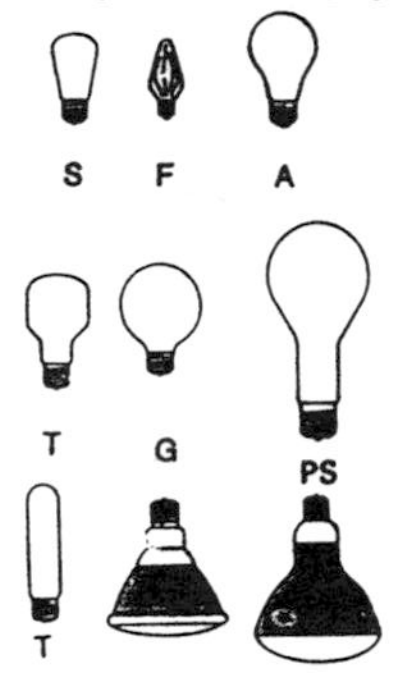


| Types | | | DISTRIBUTION: BEAM TYPE/SPREAD | DISTRIBUTION: PROJECTION RANGE: S = short = 0'-8' M = 8'-12' L = long = over 12' | DISTRIBUTION: EFFECT ON SURFACE PERPENDICULAR TO BEAM: C = circular E = elliptical | OPERATION: LUMENS (measured within primary distribution) | OPERATION: APPROX: INITIAL EFFICACY (lumens/watt) | OPERATION: LUMEN DEPRECIATION (%) | OPERATION: WATTAGE RANGE | OPERATION: BALLAST WATTAGE | OPERATION: VOLTAGE | OPERATION: AVERAGE LIFE (HRS.) | BULB DESCRIPTION: DIAMETERS (INCHES) | BULB DESCRIPTION: LENGTHS (INCHES) |
|---|---|---|---|---|---|---|---|---|---|---|---|---|---|---|
| General Service | Type S | | Broad | S | C,E | 44-80 | 8-30 | 80-90 | 6,10 | N/A | 115-125 | 1500 | 1 3/4 | 3 1/2 |
| General Service | Type A | | Broad | S-M | C,E | 125-4000 | 8.3-20 | 79-93 | 15-200 | N/A | 115-277 | 750-2500 | 1 7/8, 2 3/8, 2 5/8, 2 7/8 | 3 1/2-6 9/16 |
| General Service | Type G | | Broad | S | C | 125-35500 | 8-22 | 80-85 | 25-1000 | N/A | 115-277 | 200-800 | 3 1/8-8 | 3 7/16-8 |
| General Service | Type PS | | Broad | S-L | C,E | 2660-34000 | 13.9-22.6 | 78-90 | 150-1500 | N/A | 115-277 | 750-2500 | 3 1/8, 3 3/4, 4 3/8, 5, 6 1/2 | 6 15/16 - 13 1/16 |
| General Service | Tubular | | Linear | S-M | E | 244-800 | 8.5-27 | 80-85 | 25-75 | N/A | 115-277 | 1000,1500 | 1 5/16, 1, 1 1/4 | 3 1/8 - 17 3/4 |
| General Service | Decorative | | Point | S | C,E | | 6-10 | 80-85 | 6-60 | N/A | 115-125 | 750,1000,1500 | | 1-4 7/16 |
| Reflector | Type R-20 | Flood | 85° / 90° | S | C | 150 / 430 | 4-8 | 80-85 | 30 / 50 | N/A | 120 | 2000 | 2 1/2 | 3 15/16 |
| Reflector | Type R-30 | Spot / Flood | 30° / 130° | S / S-M | C | 400 / 610 | 5-9 | 80-85 | 75 / 175 | N/A | 120 | 2000 | 3 3/4 | 5 3/8 |
| Reflector | Type R-40 | Spot | 37° / 35° / 60° | L / L / M-L | C | 835 / 1800 / 3300 | 5-11 | 80-85 | 150 / 300 / 500 | N/A | 120 | 2000 | 5 | 6 1/2 / 6 7/8, 6 1/2, 7 1/4 / 7 1/4 |
| Reflector | Type R-40 | Flood | 110° / 115° / 120° | M / M / M | C | 1550 / 3000 / 5700 | 5-11 | 80-85 | 150 / 300 / 500 | N/A | 120 | 2000 | 5 | 6 1/2 / 6 7/8, 6 1/2, 7 1/4 / 7 1/4 |
| Ellipsoidal Reflector | Type R-52 | Flood | 90° / 110 | M-L | E | 7750, 13000 / 16300 | 15 / 17 | 80-85 | 500,756 1000 | N/A | 120 | 2000 | 6 1/2 | 11 3/4 |
| Ellipsoidal Reflector | Type RB-52 | Flood | 130° | M | E | 18900 | 18 | 80-85 | 1000 | N/A | 120 | 2000 | 6 1/2 | 12 3/4 |
| Ellipsoidal Reflector | Type ER-30 | Flood | Medium | M | C,E | 420-950 | 1.5-4.0 | 80-85 | 250 | N/A | 120 | 2000 | 6 3/8 | 6 3/8 |
| Ellipsoidal Reflector | Type ER-40 | Flood | Medium | M | C,E | 1475 | 6 | 80-85 | 250 | N/A | 120 | 2000 | 7 3/8 | 7 3/8 |
| Parabolic Aluminized Reflector | PAR-36 | Spot / Flood | Narrow / Wide | S / L | C,E | | | 80-85 | 25,50 | N/A | 120 | 2000 | | |
| Parabolic Aluminized Reflector | PAR-38 | Spot / Flood | 30° × 30° / 60° × 60° | M-L / S-M | C | 465-110 / 570-1350 | 6-9 | 80-85 | 75,100 150 | N/A | 120 | 2000 | 4 3/4 | 4 5/16, 5 5/16 |
| Parabolic Aluminized Reflector | PAR 46 | Spot / Flood | 17° × 23° / 20° × 40° | S-L | E | 1200 / 1300 | 6-8 | 80-85 | 200 | N/A | 120 | 2000 | 5.6 | 4 |
| Parabolic Aluminized Reflector | PAR-56 | Narrow Spot / Medium Flood / Wide Flood | 15° × 20° / 20° × 35° / 30° × 60° | S-L | E | 1800 / 2000 / 2100 | 6-7 | 80-85 | 300, 500 | N/A | 120 | 2000 | 7 | 5 |
| Parabolic Aluminized Reflector | PAR-64 | Narrow Flood / Medium Flood / Wide Flood | 13° × 20° / 20° × 35° / 35° × 65° | M-L | E | 3000 / 3400 / 3500 | 6-7 | 80-85 | 500, 1000 | N/A | 120 | 2000 | 8 | 6 |
| Tungsten-Halogen | | Double Ended / Single Ended | Linear | M | E | 3460-35800 / 1400-22400 | 17.3-24.5 / 17.5-25 | 93-96 | 200-1500 / -1000 | N/A | 120-277 / 12-130 | 500-4000 / 1000-3000 | 3/8, 1/2, 3/4 | 3 1/8 -10 1/16 / 2 1/4-9 1/2 |
| Tungsten-Halogen | PAR-38 | Spot / Medium Flood | 60° × 60° / 28° × 28° | L / M | C | 1800 / 2400 | 6-10 | 93-96 | 250 | N/A | 120-277 | 4000 | 5.6 | 5 5/16 |
| Tungsten-Halogen | PAR-56 | Narrow Spot / Medium Flood / Wide Flood | 15° × 32° / 20° × 42° / 34° × 66° | L | E | 4900 / 5700 / 5725 | 9-12 | 93-96 | 500 | N/A | 120-277 | 4000 | 7 | 5 |
| Tungsten-Halogen | PAR-64 | Narrow Flood / Medium Flood / Wide Flood | 14° × 31° / 22° × 45° / 45° × 72° | L | E | 8500 / 10000 / 13500 | 8.5-13.5 | 93-96 | 1000 | N/A | 120-277 | 4000 | 8 | 6 |

| Types | | | COATINGS: CLEAR | COATINGS: FROSTED | COATINGS: SILICA | COATINGS: WHITE/WHITE BOWL | COATINGS: COLORED/ENAMELED | COATINGS: SILVER BOWL/SEMI-SILVERED/REFLECTORIZED | PHOSPHOR: STANDARD | PHOSPHOR: COLOR IMPROVED | PHOSPHOR: DAYLIGHT | JACKETED/OUTDOOR/LOW-TEMP. | HEAT RESISTANT | PRISMATIC | BASE DESCRIPTION: MINIATURE/BAYONET/CANDELABRA | SCREW: MEDIUM/MED. SKIRTED | SCREW: MOGUL | END/SIDE PRONG | RECESSED: SINGLE CONTACT | RECESSED: DOUBLE CONTACT | BIPIN: MIN | BIPIN: MEDIUM | BIPIN: MOGUL |
|---|---|---|---|---|---|---|---|---|---|---|---|---|---|---|---|---|---|---|---|---|---|---|---|
| General Service | Type S | | ● | ● | | | ● | | | | | | | | | ● | | | | | | | |
| General Service | Type A | | ● | ● | ● | ● | ● | ● | | | | | | | | ● | | | | | | | |
| General Service | Type G | | ● | ● | | | ● | | | | | | | | | ● | | | | | | | |
| General Service | Type PS | | ● | ● | ● | ● | ● | ● | | | | | | | | ● | ● | | | | | | |
| General Service | Tubular | | ● | ● | | ● | ● | ● | | | | | | | | ● | | | | | | | |
| General Service | Decorative | | ● | | | | ● | | | | | | | | ● | | | | | | | | |
| Reflector | Type R-20 | Flood | | ● | | | ● | ● | | | | | | | | ● | | | | | | | |
| Reflector | Type R-30 | Spot / Flood | ● | ● | | | ● | ● | | | | | | | | ● | | | | | | | |
| Reflector | Type R-40 | Spot | ● | ● | | | ● | ● | | | | | | | | ● | ● | | | | | | |
| Reflector | Type R-40 | Flood | | ● | | | ● | ● | | | | | | | | ● | ● | | | | | | |
| Ellipsoidal Reflector | Type R-52 | Flood | | ● | | | | ● | | | | | | | | | ● | | | | | | |
| Ellipsoidal Reflector | Type RB-52 | Flood | | ● | | | | ● | | | | | | | | | ● | | | | | | |
| Ellipsoidal Reflector | Type ER-30 | Flood | | ● | | | | ● | | | | | | | | | | | | | | | |
| Ellipsoidal Reflector | Type ER-40 | Flood | | ● | | | | ● | | | | | | | | | | | | | | | |
| Parabolic Aluminized Reflector | PAR-36 | Spot / Flood | ● | | | | | ● | | | | | | | | ● | | | | | | | |
| Parabolic Aluminized Reflector | PAR-38 | Spot / Flood | ● | ● | | | | ● | | | | | | ● | | ● | | ● | | | | | |
| Parabolic Aluminized Reflector | PAR 46 | Spot / Flood | ● | | | | | ● | | | | | | | | | | | | | | | |
| Parabolic Aluminized Reflector | PAR-56 | Narrow Spot / Medium Flood / Wide Flood | ● | | | | | ● | | | | | | ● | | | | | | | | | |
| Parabolic Aluminized Reflector | PAR-64 | Narrow Flood / Medium Flood / Wide Flood | | | | | | ● | | | | | | ● | | | | | | | | | |
| Tungsten-Halogen | | Double Ended / Single Ended | | | | | | ● | | | | | | | ● | | | | ● | | | | |
| Tungsten-Halogen | PAR-38 | Spot / Medium Flood | ● | ● | | | | ● | | | | | | | | ● | | | | | | | |
| Tungsten-Halogen | PAR-56 | Narrow Spot / Medium Flood / Wide Flood | ● | | | | | ● | | | | | | ● | | | | ● | | | | | |
| Tungsten-Halogen | PAR-64 | Narrow Flood / Medium Flood / Wide Flood | ● | | | | | ● | | | | | | ● | | | | ● | | | | | |

Figure L-25

Sweet's Division
McGraw-Hill Information Systems Company

## L-34  TUNGSTEN-HALOGEN LAMPS

Halogen lamps, also called *quartz* lamps, were invented by **General Electric Corp.** in 1957. (Fig. L-24.)

They use an electronegative **"halogen gas"** (bromine, iodine, chlorine, fluorine, etc.).

Halogen vapor present in the lamp combines with particles of tungsten that have evaporated from the filament and redeposits them on the filament.

Halogen gas allows higher filament temperature.

They are available in different sizes and shapes.

Low-voltage halogen lamps require a transformer.

They can be single- or double-ended.

The types and properties of incandescent lamps and halogen lamps are given in Fig. L-25.

## L-35  FLUORESCENT LAMPS

Sir **G. G. Stokes** in 1852 discovered that fluorescence can be induced in certain substances by stimulation with ultraviolet light.

Stokes' law states: The wavelength of the fluorescent light is always greater than that of the exciting radiation with some exceptions.

**A. E. Becquerel** in 1867 constructed a fluorescent lamp and described the preparation of fluorescent tubes basically similar to those made today, Fig. L-26.

During the 1920s, high-voltage fluorescent tubes were used in advertising signs, and were called **electric discharge** lamps.

In 1937, the first low-voltage fluorescent lamps were marketed in the United States and in a short period of time replaced all the incandescent lamps in all the spaces within buildings except for residential use and specialty lighting.

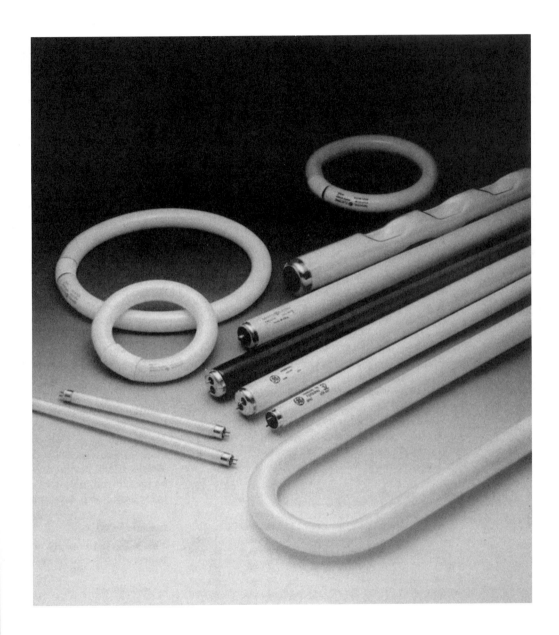

**Figure L-26   Fluorescent Lamps**

## L-36   CONSTRUCTION OF FLUORESCENT LAMPS

Fluorescent lamps (Fig. L-27) consist of the following:

a.   A cylindrical glass tube, commonly straight, circular, or U-shaped.

b.   The glass contains argon gas and low-pressure mercury vapor.

c.   Two cathodes, one on each end of the glass tube, supply power to electrons in order to start and maintain the mercury arc.

d.   On the inside, the glass tube is coated with phosphor in order to transform ultraviolet radiation into visible light.

e.   The composition of phosphor directly affects the color of light such as daylight, warm white, cool white, etc.


## L-37   OPERATION OF FLUORESCENT LAMPS

Fluorescent lamps work on the principle of gaseous discharge.

**Fluorescent lamps are favored over incandescent lamps because:**

a.   **They produce 3 to 5 times greater lumens per watt.**

b.   **They have 7 to 20 times greater lamp life.**

c.   **They produce less heat.**

d.   **They can operate at temperatures as low as –20°F.**

e.   **Lamp temperature at peak light is 104°F.**

The distribution of energy output is approximately:

**20 percent ultraviolet radiation**

**35 percent infrared radiation**

**40 percent heat radiation**

**5 percent visible light**

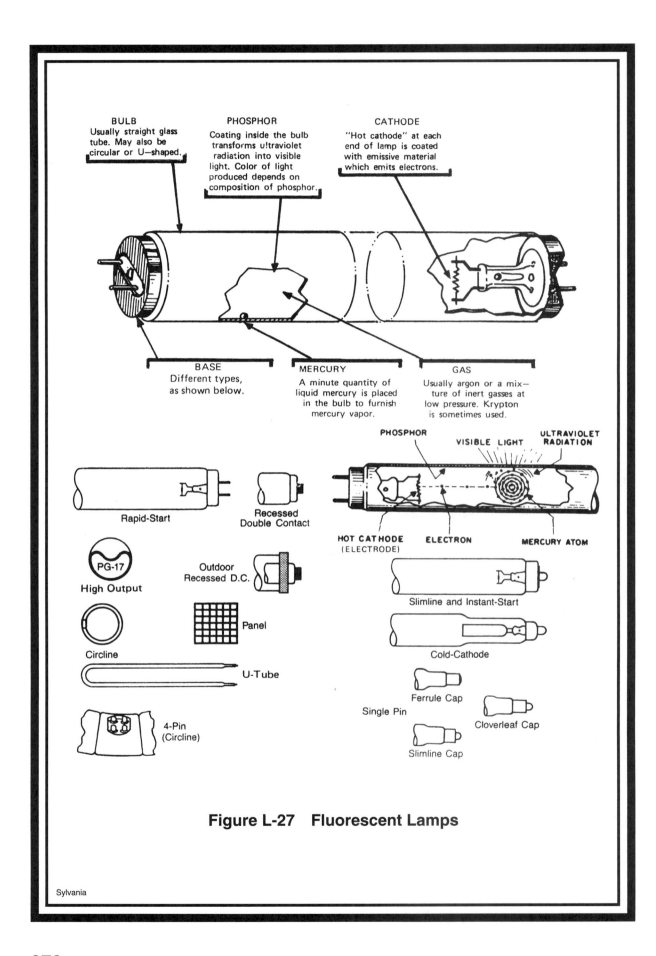

**BULB**
Usually straight glass tube. May also be circular or U—shaped.

**PHOSPHOR**
Coating inside the bulb transforms ultraviolet radiation into visible light. Color of light produced depends on composition of phosphor.

**CATHODE**
"Hot cathode" at each end of lamp is coated with emissive material which emits electrons.

**BASE**
Different types, as shown below.

**MERCURY**
A minute quantity of liquid mercury is placed in the bulb to furnish mercury vapor.

**GAS**
Usually argon or a mix— ture of inert gasses at low pressure. Krypton is sometimes used.

PHOSPHOR    VISIBLE LIGHT    ULTRAVIOLET RADIATION

HOT CATHODE (ELECTRODE)    ELECTRON    MERCURY ATOM

Rapid-Start

Recessed Double Contact

PG-17
High Output

Outdoor Recessed D.C.

Slimline and Instant-Start

Cold-Cathode

Circline

Panel

U-Tube

4-Pin (Circline)

Ferrule Cap

Single Pin

Cloverleaf Cap

Slimline Cap

## Figure L-27  Fluorescent Lamps

Sylvania

## L-38   TYPES OF FLUORESCENT LAMPS

The efficacy (lumens/watt) and color of light are affected by the selection of the phosphor compound.

The coatings of phosphor are identified by lamp names and light radiation in **kelvin (K).**

Fluorescent lighting has poor beam control, but it has an effective line of light.

The types of fluorescent lamps are:

1.   ***Daylight lamps***—6250 K

     Produce a bluish white light

     **Enhance** colors of green and blue

     **Dull** colors of orange and red

2.   ***Cool-white lamps***—4250 K

     Produce a white light

     **Enhance** colors of orange, yellow, and blue

     **Dull** and gray color of red

3.   ***Deluxe cool-white lamps***—4050 K

     Produce a white light

     **Enhance** all colors

     **Dulling** effect not noticeable

4.   ***White lamps***—3450 K

     Produce a pale yellowish white light

     **Enhance** colors of orange and yellow

     **Dull** colors of red, blue, and green

5.   ***Warm-white lamps***—3020 K

     Produce a yellowish white light

     **Enhance** colors of orange and yellow

     **Dull** colors of red, blue, and green

6.   ***Deluxe warm-white lamps***—2940 K

     Produce pinkish white light

     **Enhance** colors of red and orange

     **Gray** colors of green and blue

Information on types of properties of fluorescent lamps is given in Fig. L-28.

# LAMPS: types, properties
## FLUORESCENT, HIGH INTENSITY DISCHARGE

| Types | | DISTRIBUTION: BEAM TYPE/SPREAD | PROJECTION RANGE: S = short = 0'-8' M = 8'-12' L = long = over12' | EFFECT ON SURFACE PERPENDICULAR TO BEAM: C = circular E = elliptical | OPERATION: LUMENS (measured within primary distribution) | APPROX: INITIAL EFFICACY (lumens/watt) | LUMEN DEPRECIATION (%) | WATTAGE RANGE | BALLAST WATTAGE | VOLTAGE | AVERAGE LIFE (HRS.) | BULB DESCRIPTION: DIAMETERS (INCHES) | LENGTHS (INCHES) |
|---|---|---|---|---|---|---|---|---|---|---|---|---|---|
| Preheat | Type T5 | Linear | S | E | 85-880 | 22-63 | 67-75 | 4-13 | 2.5 | 29-95 | 6000-7500 | ⅝ | 6,9,12,21 |
| Preheat | Type T8 | Linear | S | E | 610-2255 | 22-63 | 79 | 15,30 | 4.5, 10.5 | 55-99 | 7500 | 1 | 18,36 |
| Preheat | Type T12 | Linear | S | E | 476-3185 | 22-63 | 79-85 | 14-40 | 4.5-12 | 40-101 | 7500, 9000, 20000 | 1½ | 15,18,24, 28,33,48 |
| Preheat | Type T17 | Linear | S | E | 5525-6400 | 22-63 | 82 | 90-100 | 20 | 65 | 9000 | 2⅛ | 60 |
| Preheat | Energy-saving Type T12 | Linear | S | E | 2800-2900 | 82-85 | | 34 | | 84 | 15000 | 1½ | 48 |
| Preheat | Energy-saving Type T17 | Linear | S | E | 5400-6400 | 64-78 | | 82-84 | | | 9000 | 2⅛ | 60 |
| Rapid Start | Circline Type T9 / Type T10 | Diffuse | S | C | 630-2650 | 60-71 | 72-82 | 20,22.5 / 33,41.5 | | 48,61 / 81,108 | 12000 | 1⅛ / 1¼ | 6½,8¼,ø / 12,16 ø |
| Rapid Start | U-shaped | Linear | S | E | 1480-2965 | 60-71 | 84 | 40.5-41 | 13 | 100-103 | 12000 | 1½ | 24 |
| Rapid Start | Type T10 low | Linear | S | E | 2270-3252 | 60-71 | 66 | 41 | 12-13 | 104 | 24000 | 1¼ | 48 |
| Rapid Start | Type T10 medium | | | | | | | | | | | | |
| Rapid Start | Type T10 high | | | | 4690-14000 | | | 105-205 | | 80-160 | 9000 | | 48,72,96 |
| Rapid Start | Type T12 low | Linear | S | E | 1505-3250 | 60-71 | 81-84 | 32.4,41 | 10.5-13 | 81-101 | 15,000-20,000 | 1½ | 36,48 |
| Rapid Start | Type T12 medium | | | | 1400-9200 | | 77-82 | 37-113 | | 41-153 | 9000-12000 | | 24-96 |
| Rapid Start | Type T12 high | | | | 4900-15250 | | 69-72 | .116-215 | | 84-163 | 9000 | | 48-96 |
| Rapid Start | Type PG17 | Linear | S | E | 5200-16000 | 60-71 | 69 | 116-215 | | 84-163 | 9000 | 2⅛ | 48,72,96 |
| Rapid Start | Panel | Diffuse | S | C | 3200-4800 | 60-71 | | 55-96 | | | 7500 | 1½ | 11⅝ × 11⅝ |
| Rapid Start | Energy-saving Type T12 | Linear | S | E | 1940-9100 | 46-78 | | 25-195 | | 64-137 | 10,000-20,000 | 1½ | 36,48,96 |
| Rapid Start | Energy-saving Type PG17 | Linear | S | E | 6550-14900 | 69-81 | | 95-185 | | 64-144 | 12000 | 2⅛ | 48,96 |
| Instant Start | Type T12 | Linear | S | E | 2560-3150 | 52-85 | 83 | 40.5 | 23.5-24 | 104 | 7500-12000 | 1½ | 48 |
| Instant Start | Type T17 | Linear | S | E | 1990-2940 | 52-63 | 89 | 42 | 24 | 107 | 7500-9000 | 2⅛ | 60 |
| Instant Start | Energy-saving Type T8 | Linear | S | E | 3725-3795 | 93 | | 40-41 | | | 7500 | 1 | 96 |
| Instant Start | Energy-saving Type T12 | Linear | S | E | 2550-6000 | 85-95 | | 30-60 | | 153 | 9000-12000 | 1½ | 48,96 |
| Instant Start | Slimline Type T6 | Linear | S | E | 1265-3050 | 52-63 | 76-77 | 25.5, 38.5 | 13,16 | 150,233 | 7500 | ¾ | 42,64 |
| Instant Start | Slimline Type T8 | Linear | S | E | 2100-4265 | 52-63 | 83-89 | 38-51 | 16 | 220-295 | 7500 | 1 | 72,96 |
| Instant Start | Slimline Type T12 | Linear | S | E | 1010-6365 | 52-63 | 78-91 | 21.5-75 | 16 | 53-197 | 7500-12000 | 1½ | 24-96 |
| Instant Start | Cold-Cathode | Linear | S | E | 950-3400 | 38-52 | | 26-65 | 9-19 | 240-450 | 12500-30000 | 1 | 45-93 |
| HIGH INTENSITY DISCHARGE | | | | | | | | | | | | | |
| Mercury | Type H | Broad | S-L | C,E | 580-63,000 | 30-65 | 52-89 | 40-1075 | 7-570 | 90-265 | 12000-24000 + | 3⅛ - 7 | 4 11/32 - 17⅞ |
| Metal Halide | Type M | Broad | S-L | C,E | 8300-210,000 | 69-125 | 66-75 | 100-3500 | 460-1080 | 130-270 | 2000- 20000 | 3½,4¾, 7 | 6½- 19 5/16 |
| High-Pressure Sodium | Type S | Broad | S-L | C,E | 2150-140,000 | 60-140 | 83-88 | 35-1000 | 85-1100 | 162-706 | 16000 - 24000 | 2⅛,3⅛,3½,4¾ | 4 5/16- 15 7/16 |
| Low-Pressure Sodium | | Broad | S | E | 1800-33,000 | up to 183 | 91-95 | 18-180 | | 57-240 | 10000- 20000 | 2⅛, 2⅛ | 9⅝,20, 36 |
| Self-Ballasted Mercury | | Broad | S-L | C,E | 1700-10,880 | 20-25 | | 160-250 | | 120-277 | 8000- 20000 | | 6½- 15⅜ |

| Types | | COATINGS: CLEAR | FROSTED | SILICA | WHITE/WHITE BOWL | COLORED/ENAMELED | SILVER BOWL/SEMI-SILVERED/REFLECTORIZED | PHOSPHOR: STANDARD | PHOSPHOR: COLOR IMPROVED | PHOSPHOR: DAYLIGHT | BASE DESCRIPTION: JACKETED/OUTDOOR/LOW-TEMP. | HEAT RESISTANT | PRISMATIC | MINIATURE/BAYONET/CANDELABRA | SCREW: MEDIUM/MED. SKIRTED | SCREW: MOGUL | END/SIDE PRONG | RECESSED: SINGLE CONTACT | RECESSED: DOUBLE CONTACT | BIPIN: MIN | BIPIN: MEDIUM | BIPIN: MOGUL | FOUR PIN/RECESSED BIPIN | SINGLE PIN | CAP: FERULE, SLIMLINE, CLOVERLEAF |
|---|---|---|---|---|---|---|---|---|---|---|---|---|---|---|---|---|---|---|---|---|---|---|---|---|---|
| Preheat | Type T5 | | | | | | | ● | ● | ● | | | | | | | | | | ● | | | | | |
| Preheat | Type T8 | | | | | | | ● | ● | ● | | | | | | | | | | | ● | | | | |
| Preheat | Type T12 | | | | | | | ● | ● | ● | | | | | | | | | | | ● | | | | |
| Preheat | Type T17 | | | | | | | ● | ● | ● | | | | | | | | | | | | ● | | | |
| Preheat | Energy-saving Type T12 | | | | | | | ● | ● | | | | | | | | | | | | ● | | | | |
| Preheat | Energy-saving Type T17 | | | | | | | ● | ● | ● | | | | | | | | | | | | ● | | | |
| Rapid Start | Circline Type T9 | | | | | | | ● | ● | ● | | | | | | | | | | | | | ● | | |
| Rapid Start | Circline Type T10 | | | | | | | ● | ● | ● | | | | | | | | | | | | | ● | | |
| Rapid Start | U-shaped | | | | | | | ● | ● | | | | | | | | | | | | ● | | | | |
| Rapid Start | Type T10 low | | | | | | | ● | ● | | | | | | | | | | | | ● | | | | |
| Rapid Start | Type T10 medium | | | | | | | ● | | | | | | | | | | | ● | | | | | | |
| Rapid Start | Type T10 high | | | | | | | ● | | | ● | | | | | | | | ● | | | | | | |
| Rapid Start | Type T12 low | | | | | | | ● | ● | | | | | | | | | | | | ● | | | | |
| Rapid Start | Type T12 medium | | | | | | | ● | | ● | | | | | | | | | ● | | | | | | |
| Rapid Start | Type T12 high | | | | | | | ● | | | ● | | | | | | | | ● | | | | | | |
| Rapid Start | Type PG17 | | | | | | | ● | | | | | | | | | | | ● | | | | | | |
| Rapid Start | Panel | | | | | | | ● | | | | | | | | | | | ● | | | | ● | | |
| Rapid Start | Energy-saving Type T12 | | | | | | | ● | ● | ● | | | | | | | | | ● | | ● | | | | |
| Rapid Start | Energy-saving Type PG17 | | | | | | | ● | | | | | | | | | | | ● | | | | | | |
| Instant Start | Type T12 | | | | | | | ● | ● | ● | | | | | | | | | | | ● | | | | |
| Instant Start | Type T17 | | | | | | | ● | ● | ● | | | | | | | | | | | | ● | | | |
| Instant Start | Energy-saving Type T8 | | | | | | | ● | ● | | | | | | | | | | | | | | | ● | |
| Instant Start | Energy-saving Type T12 | | | | | | | ● | ● | ● | | | | | | | | | | | | | | ● | |
| Instant Start | Slimline Type T6 | | | | | | | ● | ● | ● | | | | | | | | | | | | | | ● | |
| Instant Start | Slimline Type T8 | | | | | | | ● | | ● | | | | | | | | | | | | | | ● | |
| Instant Start | Slimline Type T12 | | | | | | | ● | | ● | | | | | | | | | | | | | | ● | |
| Instant Start | Cold-Cathode | | | | | | | | ● | ● | | | | | | | | | | | | | | | ● |
| HIGH INTENSITY DISCHARGE | | | | | | | | | | | | | | | | | | | | | | | | | |
| Mercury | Type H | ● | | | | | ● | ● | ● | | | ● | | | ● | ● | | | | | | | | | |
| Metal Halide | Type M | | | | | | ● | ● | ● | | | ● | | | | ● | | ● | | | | | | | |
| High-Pressure Sodium | Type S | | | | | | ● | ● | ● | ● | | | | | ● | ● | | ● | | | | | | | |
| Low-Pressure Sodium | | ● | | | | | | | | | | | | | | | | | | | | | | ● | |
| Self-Ballasted Mercury | | ● | | | | | ● | ● | ● | | | ● | | | ● | ● | | | | | | | | | |

Sweet's Division
McGraw-Hill Information Systems Company

Sweet's Division
McGraw-Hill Information Systems Company

**Figure L-28**

**L-39    FLUORESCENT LAMPS STARTERS**

There are three methods to generate the *arc* in fluorescent lamps, and they are identified by the type of lamps:

1.    *Preheat lamps*
2.    *Instant-start lamps*
3.    *Rapid-start lamps*

**L-40    PREHEAT LAMPS**

These have starters and ballasts, which raise the temperature of the filament cathode to produce electron emission.

The starter operates manually or automatically.

Starter types are thermal switch, glow-switch and cutout starters.

**From initial power to full light output takes 2 to 4 seconds.**

The application of this lamp is limited to 30 watts and 36-inch lamp length.

**L-41    INSTANT-START LAMPS**

Initially, 400 to 1000 volts are used to start an electrode to strike the arc and **produce full light instantly.**

In this type of lamp, only a single-pin base is required.

They have longer life, lower efficacy, and lower output per foot of length than preheat lamps.

**L-42    RAPID-START LAMPS**

In this type of lamp, ballasts operate without starters.

They require a grounding strip for the full length of the lamp.

They require double contact bases.

The ballasts heat the cathodes when the lamp is in operation; therefore, additional wattage is used.

**They require 1 to 2 seconds to reach full light output.**

**Figure L-29   Compact Fluorescent Lamps**

**L-43   VOLTAGE FOR BALLAST**

Operating voltage for ballasts are as follows:

a.   110 to 125 V on 120-V circuits

b.   200 to 215 V on 208-V circuits

c.   250 to 290 V on 277-V circuits


**L-44   COMPACT FLUORESCENT LAMPS**

They are also called *miniature fluorescent lamps.* (Fig. L-29.)

These lamps work on the same principle as regular fluorescent lamps and offer smaller size, lower cost to operate, and easier installation.

There are two types of compact lamps:

1.   The lamp is designed to be used in fluorescent lampholders and auxiliary ballasts.

2.   The lamp is designed to allow installation in ordinary incandescent sockets.

**They have a long life, approximately 10 times longer than incandescent lamps.**

> **Lamp wattages are low: 5, 7, 9, 13, and 18 W.**
>
> **Illumination output is: 250 to 1250 lumens.**
>
> **Estimated life range: 7500 to 10,000 hours.**

The use of compact lamps is limited to indoor applications.


**L-45   HIGH-INTENSITY DISCHARGE (HID) LAMPS**

In fluorescent lamps the arc produces only the ionization of mercury vapor. (Fig. L-30.)

In **HID** lamps the arc continues throughout the lamp operation, and it is the actual source of light radiation.

Ballasts are used for both starting and operation, using high-input wattage.

They have outer-glass bulbs which protect the inner arc tubes.

**HID** lamps' temperature at peak output is 300 to 400°F.

The bulb types are standard, reflector, tubular, globe, and elliptical.

**Figure L-30   High Intensity Discharge Lamps**

There are four types of **HID** lamps:

1. *Mercury HID lamps*
2. *Metal halide HID lamps*
3. *High-pressure sodium (HPS) HID lamps*
4. *Low-pressure sodium (LPS) HID lamps*

## L-46  MERCURY HID LAMPS

This lamp has a third electrode in the arc tube which is used to ionize *argon* **gas** in order to raise the temperature and pressure within the tube for vaporizing the *mercury* for arc generation. (Figs. L-28 and L-30.)

In addition to argon gas, the arc tubes may contain *neon* and *krypton* gases.

*Nitrogen* is sealed between the arc tube and the outer bulb in order to stop *oxidation* of the internal parts.

In some of the **HID** lamps, the outer bulb is coated on the inside with *phosphor* for improving the color.

This type of lamp converts approximately 12 percent of input energy into visible light radiation.

They are usually used indoors in commercial and industrial buildings.

They are used outdoors for monument lighting, landscape lighting, and flood lighting.

They can operate below −20°F with special ballasts.

Their operation becomes affected if they are used above 105°F.

*Mercury HID lamps "clear"*—5710 K

Produce greenish-bluish white light

**Enhance** colors yellow, blue, and green

**Gray** colors red and orange

## L-47 METAL HALIDE HID LAMPS

These lamps are smaller than similar-wattage mercury **HID** lamps. (Figs. L-28 and L-30.)

Their construction and operation is almost identical to mercury HID lamps, but they are distinguished by the addition of *metal halides:*

> Sodium
>
> Thallium and indium iodides
>
> Sodium and scandium iodides
>
> Dysprosium and thallium iodides

Their lumens per watt are twice that of the mercury **HID** lamps.

These lamps convert approximately 21 percent of wattage input into visible light.

They are used for floodlighting, sports lighting, parking area lighting, and sign lighting.

*Metal halide HID lamps "clear"*—3720 to 4200 K

> Produce greenish white light
>
> **Enhance** the colors yellow, green, and blue
>
> **Dull** color red

## L-48 HIGH-PRESSURE SODIUM (HPS) HID LAMPS

In this lamp, light is produced by passing the arc through a combination of *sodium, mercury,* and *xenon* for starting. (Figs. L-28 and L-30.)

The arc tube is made of *polycrystalline alumina,* which is contained in a vacuum and is protected by the outer tube.

These lamps are more efficient than mercury **HID** lamps and metal halide **HID** lamps. They produce 26 percent of input wattage to visible light.

They can operate under –30°F with special ballasts.

These lamps require 4 to 8 minutes to warm up.

Bulb temperatures can be as high as 400°F, and the base temperature up to 210°F.

They are used for roadway lighting, sports lighting, and other outdoor lighting.

### High-pressure sodium (HPS) HID lamps "clear"—2100 K

Produce golden-white or yellowish light

**Enhance** colors yellow, green, and orange

**Dull** or gray colors blue and red

## L-49   LOW-PRESSURE SODIUM (LPS) LAMPS

They produce light by an electric charge through *sodium* vapor. (Figs. L-28 and L-30.)

A *neon* starting gas is ionized, which produces a red glow while warming the sodium. When the sodium is vaporized, it produces yellow light. Approximately 35 percent of the input wattage is converted to visible light.

This lamp requires 7 to 15 minutes to reach full light output, with a temperature of 500°F.

Its color temperature is 1700 K and produces a yellow/orange or brownish color light.

It is used for outdoor lighting.

## L-50   NEON LAMPS

Neon lamps are custom-fabricated, and are used for continuous lines of light. (Fig. L-31.)

They have many applications, such as lighting arches, stairs, and signs.

They are the **cold-cathode** type, with tubes ½ to 1 inch and lengths up to 120 feet, which can be connected to only two leads.

The glass tubes are heated to form any shape desired.

**The lamp color is based on the gas contained in the lamp.**

| | | |
|---|---|---|
| **Neon** | = | orange to red colors |
| **Argon and mercury** | = | blue color |
| **Helium** | = | pinkish white |

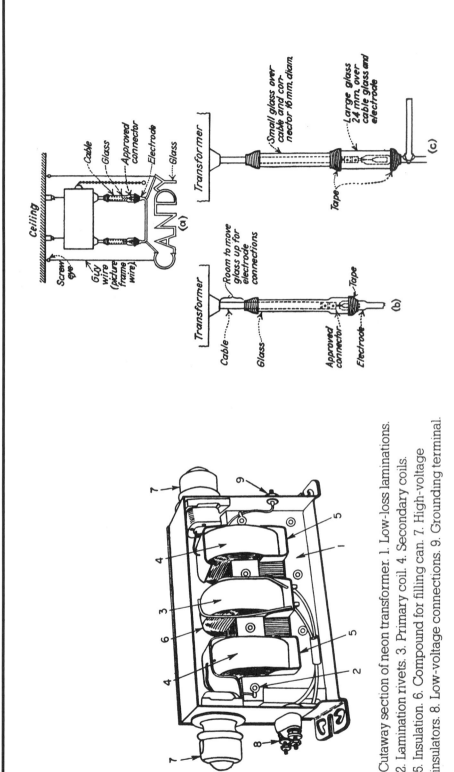

Cutaway section of neon transformer. 1. Low-loss laminations. 2. Lamination rivets. 3. Primary coil. 4. Secondary coils. 5. Insulation. 6. Compound for filling can. 7. High-voltage insulators. 8. Low-voltage connections. 9. Grounding terminal.

(a) Hanging transformer and skeleton sign. The guy wires should be completely insulated when attached to the neon tubing. Note the method of protecting electrodes and cable with glass tubing. (b) Detail of glass sleeve for protecting electrode and cable. (c) Two pieces of tubing used in place of the one shown in (b).

## Figure L-31   Neon Lamps

Acme Electric Corporation

Also, glass-tube coloring or coating will provide additional colors.

Altogether, approximately 25 colors can be produced.

A 1-inch-diameter tube produces 300 lumens per foot.

Lamp life is up to 25,000 hours.

Loose connections cause a flickering effect.

A step-up transformer is required for operation of these lamps, which need 6000 to 10,000 volts or more.

| TYPE OF ACTIVITY | FOOTCANDLES |
|---|---|
| **Assembly** | |
| Art galleries | 30–100 |
| Auditorium/assembly spaces | 15–30 |
| **Banks** | |
| Lobby | 50 |
| Customer area | 70 |
| Teller stations | 150 |
| **Churches & Synagogues** | |
| Altar, Ark | 100 |
| Pews | 20 |
| Pulpit (supplementary) | 50 |
| **Hospitals** | |
| Corridors, toilets, waiting rooms | 20 |
| Patient rooms | |
| General | 20 |
| Supplementary for reading | 30 |
| Supplementary for examination | 100 |
| Recovery rooms | 30 |
| Lab, exam, treatment rooms | |
| General lighting | 50 |
| Closework and examining table | 100 |
| Autopsy | |
| General lighting | 100 |
| Supplementary lighting | 1,000 |
| Emergency rooms | |
| General lighting | 100 |
| Supplementary lighting | 2,000 |
| Surgery | |
| General lighting | 200 |
| Supplementary on table | 2,500 |
| **Hotels and Motels** | |
| Bedrooms | 30 |
| Corridors, elevators, stairs | 20 |
| Entrance foyer | 30 |
| **Libraries** | |
| Reading rooms and carrells | 70 |
| Stacks | 30 |
| Catalogs, card files | 70 |

| TYPE OF ACTIVITY | FOOTCANDLES |
|---|---|
| **Offices** | |
| Designing, detailed drafting | 200 |
| Accounting, bookkeeping, business machine operation | 150 |
| Regular office work | 100 |
| **Post Offices** | |
| Lobby, on tables | 30 |
| Sorting, mailing | 100 |
| **Restaurants** | |
| Dining areas | |
| Cashier | 50 |
| Light environment | 10–30 |
| Kitchen | 70 |
| **Schools** | |
| Regular classrooms | 70 |
| Chalkboards | 150 |
| Drafting rooms | 100 |
| Laboratories | 70–150 |
| Manual arts | 100 |
| Sewing rooms | 150 |
| Corridors and stairs | 20 |
| Gymnasiums | 30–70 |
| **Stores** | |
| Circulation areas | 30 |
| Merchandising areas | |
| Service stores | 100 |
| Self-service stores | 200 |
| Showcases and wall cases | 200 |
| **Theaters** | |
| During intermission | 10 |
| Foyer | 10 |
| Entrance lobby | 30 |
| **Exteriors** | |
| Building security | 10–30 |
| Parking | |
| Self-parking | 20 |
| Attendant parking | 20 |
| Shopping centers | 30 |
| Floodlighting | 50–150 |
| Bulletins and poster panels | 100–150 |

**Figure L-32   Suggested Illumination Levels**

# LIGHTING DESIGN AND CALCULATION

**L-51    SYSTÈME INTERNATIONAL (SI)**

The metric system of measurements originated in France and was adopted there in 1799.
It received acceptance by many nations, and its use was permitted in the United States.
Seventeen countries including the U.S. signed the **"Treaty of the Meter"** in 1875. In 1960,
the metric system was revised, and it was called **Systeme International d'Unites (SI).**

**L-52    LUMEN (lm)**

Lumen is a unit of measurement for the flow of light output from a source.

**Lumen is the unit of *luminous flux* or *quantity of light* ($Q$) and is measured in hours (h).**

$$\text{Lumen hour} = \text{lmh} = Q$$

**L-53    ILLUMINATION or QUANTITY OF LIGHT**

This is the best result of light falling on a surface and is measured in ***footcandles* (fc).**

**L-54    FOOTCANDLES (fc)**

Footcandle is the unit used to measure the density of luminous flux or quantity of light.

**When one lumen spreads over an area of 1 square foot, it produces an illumination of
1 footcandle.**

$$\text{Footcandles (fc)} = \frac{\text{lumen (lm)}}{\text{ft}^2}$$

**When 1 lumen spreads over an area of 1 square meter (SI) it produces an illumination
of 1 *lux* (lx).**

$$\text{Lux (lx)} = \frac{\text{lumen (lm)}}{\text{m}^2}$$

$$\text{Lux (lx)} = \text{footcandle (fc)} \times 10.75$$

Recommended illumination levels for different types of spaces are given in Fig. L-32.

**L-55** **LUMINANCE (FL)**

Luminance is also called *brightness,* and is the result of light that is reflected, transmitted, or emitted from a source of light or surface. It is measured in *footlamberts* **(FL).**

$$\text{Luminance} = \text{FL} = \text{fc} \times \text{reflectance factor}$$

$$\text{FL} = \frac{\text{lumen (lm)}}{\text{area (in square feet)}} \times \text{RF}$$

**L-56** **REFLECTANCE FACTOR (RF)**

Also called *reflection factor,* it is the ratio of light reflected by a **luminaire** and the light produced by the **luminaire.**

The reflectance factor is supplied by the manufacturers of the luminaires (fixtures).

**L-57** **COEFFICIENT OF UTILIZATION (CU)**

The coefficient of utilization is the ratio between the **lumen (lm)** reaching the working plane (table, desk, floor, etc.) and the **lumen** generated from the lamps(s) inside the luminaire.

**The CU varies between 0.5 and 0.8 in most cases.**

**L-58** **LUMINOUS INTENSITY or CANDLEPOWER**

The amount of illumination emitted by a lamp in a certain direction is called *luminous intensity* (I) or *candlepower* (cd).

The unit of measurement is called *candela* (cd). Some designers use (cp):

$$\text{cd} = \text{cp}$$

**L-59** **EFFICACY**

Efficacy is the ratio of lumens produced by a light and the amount of wattage used to generate the lumens.

$$\text{Efficacy} = \frac{\text{lumens (lm)}}{\text{watts (W)}}$$

**L-60     LIGHTING CALCULATION**

There are two types of calculations used to design the lighting systems for the spaces within the building:

1.     *Local lighting calculation*
2.     *General lighting calculation*

The calculations of lighting for both systems are *simple,* requiring only a basic mathematical knowledge.

We architects are capable of designing and calculating a proper lighting system for all the spaces within our building(s)

In fact, the reflected ceiling p an designed and calculated by us will enhance the overall environment of the spaces.

**L-61     PROCEDURE FOR LOCAL LIGHTING CALCULATION**

A *light* fixture (luminaire) is located at a distance above the *work plane* (W.P.) such as table, desk, etc., and we want to know how much *light* (illuminance) is falling on the top of work plane.

The method of calculation commonly used is called the *point-by-point method.*

There are three conditions that may affect the illumination on a work plane:

1.     **Only one light is installed directly above the working plane.**
2.     **Only one light is installed away from and above the working plane.**
3.     **Two or more lights are installed directly above or away from the working plane.**

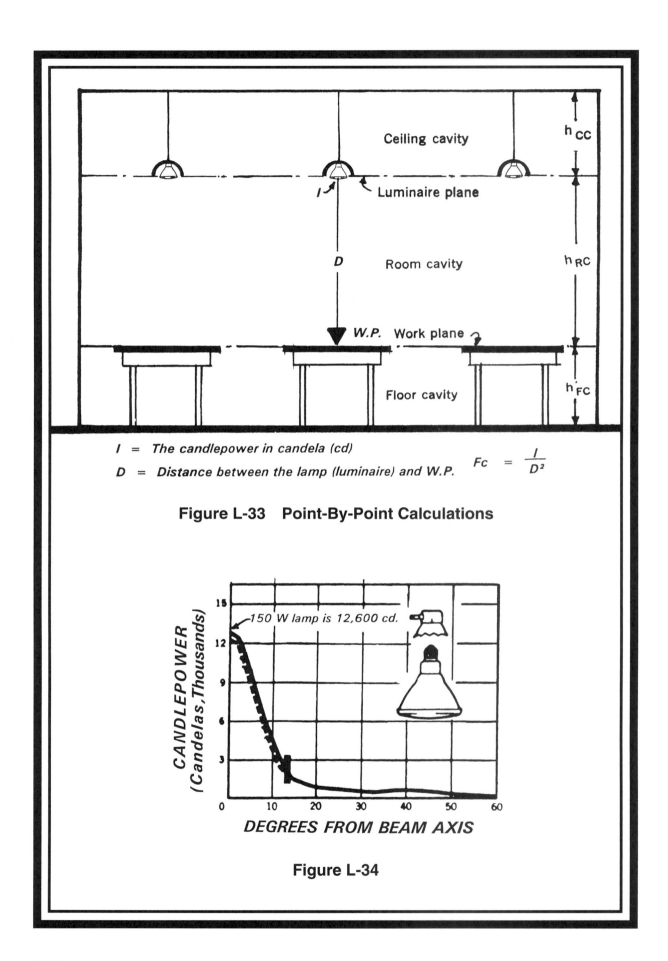

$I$ = The candlepower in candela (cd)

$D$ = Distance between the lamp (luminaire) and W.P.

$$Fc = \frac{I}{D^2}$$

**Figure L-33    Point-By-Point Calculations**

150 W lamp is 12,600 cd.

CANDLEPOWER (Candelas, Thousands)

DEGREES FROM BEAM AXIS

**Figure L-34**

# LOCAL LIGHTING CALCULATION

**L-62**  **CALCULATION FOR ONE LIGHT DIRECTLY ABOVE W.P.**

**Design criteria:**

**Step 1.** Draw a cross section of the **W.P.** and locate the light and its distance from **W.P.**

**Step 2.** Use *only* a candlepower distribution chart which is supplied by the luminaire manufacturer to obtain the value of **candela (cd).**

**Step 3.** Use the following formula to obtain footcandles on the **W.P.**

$$fc = \frac{I}{D^2}$$

where   W.P. = working plane

fc = footcandles

$I$ = candlepower in candela (cd)

$D$ = distance between the lamp (luminaire) and W.P.

## *Example*

A 150-W lamp is located 7 feet directly above the **W.P.** (Fig. L-33) What are the footcandles' illumination on the W.P.?

## *Solution*

**Step 1.** The section has been drawn and is shown in Fig. L-33.

**Step 2.** The chart shown in Fig. L-34 indicates the candlepower for 150-W lamp is 12,600 cd.

> **Note:** This chart is used *only* for demonstration. Use the proper chart supplied by a luminaire manufacturer.

**Step 3.**

$$fc = \frac{I}{D^2}$$

$$fc = \frac{12,600 \text{ cd}}{(7)^2} = 257.15 \text{ footcandles of illumination on W.P.}$$

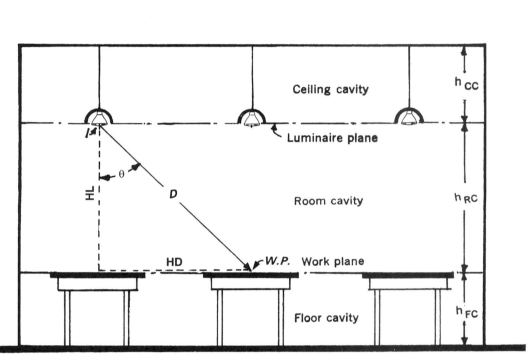

$I$ = *The candlepower in candela (cd)*

$D$ = *Distance between the lamp (luminaire) and W.P.*

$$Fc = \frac{I}{D^2}\ (\cos\ \theta)$$

**Figure L-35   Point-By-Point Calculations**

(HD)

Horizontal Distance from Axis of Light Source—feet

| HL \ HD | 0 | 1 | 2 | 3 | 4 | 5 | 6 | 7 | 8 | 9 | 10 | 11 | 12 | 13 | 14 | 15 |
|---|---|---|---|---|---|---|---|---|---|---|---|---|---|---|---|---|
| 2 | 0° 25.000 | 27° 17.850 | 45° 8.850 | 56° 4.275 | 63° 2.245 | 68° 1.298 | 71° 0.802 | 74° 0.528 | 76° 0.355 | 78° 0.255 | 79° 0.190 | 80° 0.142 | 81° 0.113 | 81° 0.090 | 82° 0.070 | 82° 0.058 |
| 3 | 0° 11.110 | 18° 9.500 | 34° 6.400 | 45° 3.933 | 53° 2.400 | 59° 1.522 | 63° 1.000 | 67° 0.680 | 69° 0.477 | 72° 0.356 | 73° 0.264 | 75° 0.205 | 76° 0.161 | 77° 0.126 | 78° 0.100 | 79° 0.084 |
| 4 | 0°0' 6.250 | 14° 5.707 | 27° 4.472 | 37° 3.200 | 45° 2.210 | 51° 1.524 | 56° 1.066 | 60° 0.764 | 63° 0.559 | 66° 0.419 | 68° 0.320 | 70° 0.249 | 72° 0.198 | 73° 0.159 | 74° 0.130 | 75° 0.107 |
| 5 | 0°0' 4.000 | 11° 3.771 | 22° 3.202 | 31° 2.522 | 39° 1.904 | 45° 1.414 | 50° 1.050 | 54° 0.785 | 58° 0.595 | 61° 0.458 | 63° 0.358 | 66° 0.283 | 67° 0.228 | 69° 0.185 | 70° 0.152 | 72° 0.126 |
| 6 | 0°0' 2.778 | 9° 2.673 | 18° 2.372 | 27° 1.987 | 34° 1.600 | 40° 1.260 | 45° 0.982 | 49° 0.766 | 53° 0.600 | 56° 0.474 | 59° 0.378 | 61° 0.305 | 63° 0.249 | 66° 0.205 | 67° 0.170 | 68° 0.142 |
| 7 | 0°0' 2.041 | 8° 1.980 | 16° 1.814 | 23° 1.585 | 30° 1.336 | 36° 1.100 | 41° 0.893 | 45° 0.722 | 49° 0.583 | 52° 0.473 | 55° 0.385 | 58° 0.316 | 60° 0.261 | 62° 0.218 | 63° 0.183 | 65° 0.154 |
| 8 | 0°0' 1.563 | 7° 1.527 | 14° 1.427 | 21° 1.283 | 27° 1.118 | 32° 0.953 | 37° 0.800 | 41° 0.666 | 45° 0.552 | 48° 0.458 | 51° 0.381 | 54° 0.318 | 56° 0.267 | 58° 0.225 | 60° 0.191 | 62° 0.163 |
| 9 | 0°0' 1.235 | 6° 1.212 | 13° 1.148 | 18° 1.054 | 24° 0.943 | 29° 0.825 | 34° 0.711 | 38° 0.607 | 42° 0.515 | 45° 0.437 | 48° 0.370 | 51° 0.314 | 53° 0.267 | 55° 0.228 | 57° 0.196 | 59° 0.168 |
| 10 | 0°0' 1.000 | 5°43' 0.985 | 11° 0.943 | 17° 0.879 | 22° 0.801 | 27° 0.716 | 31° 0.631 | 35° 0.550 | 39° 0.476 | 42° 0.411 | 45° 0.354 | 48° 0.305 | 50° 0.263 | 52° 0.227 | 54° 0.196 | 56° 0.171 |
| 11 | 0°0' 0.826 | 5°12' 0.816 | 10° 0.787 | 15° 0.742 | 20° 0.686 | 24° 0.623 | 29° 0.559 | 32° 0.496 | 36° 0.437 | 39° 0.383 | 42° 0.335 | 45° 0.292 | 48° 0.255 | 50° 0.223 | 52° 0.195 | 54° 0.171 |
| 12 | 0°0' 0.694 | 4°46' 0.687 | 9° 0.668 | 14° 0.634 | 18° 0.593 | 23° 0.546 | 27° 0.497 | 30° 0.448 | 34° 0.400 | 37° 0.356 | 40° 0.315 | 43° 0.278 | 45° 0.246 | 47° 0.217 | 49° 0.191 | 51° 0.169 |

Height of Light Source Above Surface—Feet

(HL)

**Figure L-36   Point-by-Point Calculation Table**

**L-63    CALCULATION FOR ONE LIGHT AWAY FROM AND ABOVE THE W.P.**

**Design criteria:**

**A.** Draw a cross section of the space showing W.P. and location of the light.

**B.** By using Fig. L-35 as reference, draw lines indicating the horizontal distance (HD), height of the light (HL), and the distance (D).

**Now you can solve the problem in two different ways:**

**1. By using the formula:**

Step 1. Find Cos θ.

$$\cos \theta = \frac{HL}{D}$$

Step 2. Use the following formula to obtain (fc) on WP

$$fc = \frac{I}{D^2} \times \cos \theta$$

I = candlepower in candela (cd), see Fig. L-34.

I is given for a 150 W lamp as an example only. Use the candlepower distribution chart which is supplied by your lamp manufacturer.

*Example*

A cross section of a drafting room is shown in Fig. L-35. A 150 W lamp is located 7 ft above WP (HL), and 8 ft away from WP (HD), the distance (D) is 11 ft. What is the footcandle (fc) illumination on the WP?

*Solution*

Step 1. Find Cos θ.

$$\cos \theta = \frac{HL}{D}$$

$$\cos \theta = \frac{7 \text{ ft}}{11 \text{ ft}} = 0.636$$

Step 2. Find footcandle (fc) on work plane

From Fig. L-36 for HL = 7 ft and HD = 8 ft. θ = 49°.

Using Fig. L-34 for degrees from beam axis of 49°, the candlepower cd is approximately 800 cd;

290

Therefore $\qquad fc = \dfrac{I}{D^2} \times \cos\theta$

$$fc = \dfrac{800}{(11)^2} \times 0.636 = 4.2 \text{ fc of illumination on WP}$$

## 2. By using point-by-point calculation table

### Example

Using Fig. L-35 as reference, if HL = 10 ft, HD = 12 ft and candlepower = 2000 cd. What is the footcandle (fc) on work plane?

### Solution

Using Fig. L-36 for HL = 10 ft and HD = 12 ft multiplier = 0.263 and $\theta$ = 50°.

Multiplier given is for 100 candelas (cd).

Therefore $\qquad fc = \dfrac{I}{100} \times \textbf{multiplier}$

$$fc = \dfrac{2000}{100} \times 0.263 = 5.26 \text{ fc of illumination on WP}$$

## L-64    CALCULATION FOR TWO OR MORE LIGHTS DIRECTLY ABOVE OR AWAY FROM W.P.

**Abney's law states:**

> **Light arriving at a surface is the sum of the light arriving from all sources to which the surface is exposed.**

All lights directly above the W.P. or above and away from the W.P. should be calculated individually by following L-62 and L-63.

Total footcandles for all lights involved should be added in order to obtain the footcandle illumination on the working plane.

# GENERAL LIGHTING CALCULATION

**L-65    PROCEDURE FOR GENERAL LIGHTING CALCULATION**

The design and calculation of general lighting for all types of buildings regardless of size:

1. **Requires basic mathematical knowledge**

2. **Is simple to design and calculate**

3. **Consumes very little time**

4. **Is rewarding as an accomplishment in both self-satisfaction and financial means**

This system of lighting is used for rooms or spaces within the building which require fairly **uniform lighting** on the work plane (W.P.), such as general office spaces, offices, classrooms, and laboratories, etc.

**In general lighting systems, there are two cases which are common in all buildings:**

1. **When all or most of the rooms and spaces within a building have:**

   *a.*    The same type of fixtures

   *b.*    The same height for work planes (tables, desks, etc.)

   *c.*    The same method of installation for fixtures (recessed in the ceiling, surface-mounted on the ceiling or hung from the ceiling)

   *d.*    The same ceiling height in all rooms and spaces

   All we have to do is simply find out for how many square feet **one fixture** will provide the required footcandles (req. fc) on the work plane. We divide the total square feet of each room or space by (req. fc) and obtain the number of **fixtures** required in each room or space.

2. **When some of the rooms or spaces within a building have:**

   *a.*    Different types of fixtures

   *b.*    Different heights for work planes

   *c.*    Different methods of installation for fixtures

   *d.*    Different ceiling heights in rooms or spaces

   In this case we group the rooms or spaces with the same requirements together, and design and calculate the lighting requirements for each group separately, as stated in L-66.

**ZONAL METHOD OF CALCULATION**

It is also know as *lumen method of calculation.*

This method is commonly used for the design and calculation of general lighting systems and is based on several **precalculated tables and charts** which are given in the **IES** *Lighting Handbook* and are used in this book with the permission of the IES of North America.

The zonal method of calculation is a procedure for determining the **average footcandles** (**fc**) illumination intensity requirement on the **work plane** in a room or a space. This method is based on

**1 lumen lm falling on one sq. ft. = 1 fc**

Therefore,

$$\text{fc} = \frac{\text{lm}}{\text{A}} \quad \text{or} \quad \text{A} = \frac{\text{lm}}{\text{fc}}$$

where   A = area in square feet      lm = lamp lm

The ratio between the lm generated from the luminaire and the lm reaching the work plane is called **coefficient of utilization** (CU).

Then

$$\text{A} = \frac{\text{Lamps lm} \times \text{CU}}{\text{fc}}$$

There is one other factor affecting this formula: the light loss factor (LLF).

Therefore,

$$\text{A} = \frac{\text{lamps lm} \times \text{CU} \times \text{LLF}}{\text{fc}}$$

Lamp lm = number of lamps in fixture × lm per lamp

**The final formula will be**

$$\text{A} = \frac{\text{no. of lamps/fixture} \times \text{lm/lamp} \times \text{CU} \times \text{LLF}}{\text{fc}}$$

**Known factors in the formula**

1.   *fc—footcandle.*   Footcandle requirements for all types of activity are given in Fig. L-32.

2.   **Number of lamps/fixture.**   Type of fixtures and number of lamps per fixture are given as reference only in Fig. L-37.

You have to choose the type of fixtures you wish to use in your building. You can find this information in any lighting catalog provided by lighting manufacturers. Use the latest issue.

3.    ***lm/lamp.***    Lumen per lamp is given in Fig. L-28. They are also available in lighting catalogs.

### Unknown factors in the formula

1.    ***Coefficient of utilization (CU)***

2.    ***Light loss factor (LLF)***


**L-67    COEFFICIENT OF UTILIZATION (CU)**

The coefficient of utilization is directly related to two factors:

a.    ***Reflectance in the room***

b.    ***Cavity ratio***


**L-68    REFLECTANCE IN THE ROOM**

We have to assume the initial or base reflectance for the **"ceiling, floor, and walls."**

You can determine the reflectance according to color and texture in the following scale:

| Black and matte | } | 0%, then 10, 20, 30, 40, 50, 60, 70, 80, and 90% | } | White and glossy |
|---|---|---|---|---|

### Assumed reflectance commonly used

| | | | |
|---|---|---|---|
| ***Ceiling*** | White acoustical tile | Use 80% | |
| ***Floor*** | Dark color use 10% | Light color use 30% | |
| ***Walls*** | White color use 70% | Other colors use 50% | |

**CAVITY RATIO FIG L-38**

There are two methods used for determining the cavity ratio.

1. By using the table (Fig. L-39)
2. By calculation

**L-69    CAVITY RATIO USING TABLE**

The zonal cavity method of calculation is based on the concept of dividing the room into three zones (Fig. L-38):

> Ceiling cavity
>
> Room cavity
>
> Floor cavity

And since each cavity has a height, they are called:

> **Height of ceiling cavity    hcc** ⎫
>
> **Height of room cavity    hrc**  ⎬ in feet
>
> **Height of floor cavity    hfc**   ⎭

The cavity ratio for all sizes of rooms and spaces is given in Fig. L-39, courtesy of IES of North America. The following example shows how to use the table:

*Example*

A room has a width of 20 feet and a length of 30 feet. If the cavity depths are hcc = 2 feet, hrc = 6 feet, and hfc = 2.5 feet, determine the following:

*a.* Ceiling cavity ratio, CCR

*b.* Room cavity ratio, RCR

*c.* Floor cavity ratio, FCR

*Solution*

Using Fig. L-39:

*a.* Ceiling cavity ratio    CCR = 0.8    for W = 20 ft, L = 30 ft, and CC = 2 ft

*b.* Room cavity ratio    RCR = 2.5    for W = 20 ft, L = 30 ft, and RC = 6 ft

*c.* Floor cavity ratio    FCR = 1.0    for W = 20 ft, L = 30 ft, and FC = 2.5 ft

## L-70  CAVITY RATIO BY CALCULATION

If the size of the rooms or spaces and height (depth) of the cavity falls between the numbers given in the table, it is easier and more accurate to calculate the cavity ratio than it is to use the interpolation of the table.

**For square or rectangular rooms or spaces:**

Ceiling cavity ratio

$$\textbf{CCR} = \textbf{5 hcc} \left( \frac{\textbf{L} + \textbf{W}}{\textbf{L} \times \textbf{W}} \right)$$

Room cavity ratio

$$\textbf{RCR} = \textbf{5 hrc} \left( \frac{\textbf{L} + \textbf{W}}{\textbf{L} \times \textbf{W}} \right)$$

Floor cavity ratio

$$\textbf{FCR} = \textbf{5 hfc} \left( \frac{\textbf{L} + \textbf{W}}{\textbf{L} \times \textbf{W}} \right)$$

where   $L$ = length of the room   $W$ = width of the room

### Example

A room has a width of 20 feet and a length of 30 feet. If the cavity depths are hcc = 2 feet, hrc = 6 feet, and hfc = 2.5 feet, determine the following:

a.    Ceiling cavity ratio, CCR
b.    Room cavity ratio, RCR
c.    Floor cavity ratio, FCR

### Solution

$$\frac{L + W}{L \times W} = \frac{30 + 20}{30 \times 20} = \frac{50}{600}$$

a.    $CCR = 5 \times 2 \left( \dfrac{50}{600} \right) = 0.8$

b.    $RCR = 5 \times 6 \left( \dfrac{50}{600} \right) = 2.4$

c.    $FCR = 5 \times 2.5 \left( \dfrac{50}{600} \right) = 1.0$

**For circular rooms or spaces:**

Ceiling cavity ratio $\quad$ $\mathbf{CCR} = \dfrac{5\,hcc}{r}$

Room cavity ratio $\quad$ $\mathbf{RCR} = \dfrac{5\,hrc}{r}$

Floor cavity ratio $\quad$ $\mathbf{FCR} = \dfrac{5\,hfc}{r}$

where $\quad r$ = radius of room or space

### Example

A room has a radius of 10 feet. If the cavity depths are hcc = 2 feet, hrc = 6 feet, and hfc = 2.5 feet, determine the following:

a. Ceiling cavity ratio, CCR

b. Room cavity ratio, RCR

c. Floor cavity ratio, FCR

### Solution

a. $\quad CCR = \dfrac{5 \times 2}{10} = 1.0$

b. $\quad RCR = \dfrac{5 \times 6}{10} = 3.0$

c. $\quad FCR = \dfrac{5 \times 2.5}{10} = 1.2$

### L-71 $\quad$ LIGHT LOSS FACTOR (LLF)

Light loss factor is also known as *maintenance factor* (MF).

Light loss factor is a combination of two factors:

1. **Lamp lumen depreciation (LLD)**

2. **Luminaire dirt depreciation (LDD)**

Therefore, $\quad$ **LLF = LLD $\times$ LDD**

**L-72    LAMP LUMEN DEPRECIATION (LLD or LLF)**

The initial lumens of the light bulbs will drop by certain percentages after a short period of time. This is also called **light loss factor (LLF).**

For example, a fluorescent light bulb's lumens decrease by 10 percent after the first 100 hours of use.

You may use the **"Lumen Depreciation %"** column in Figs. L-25 and L-28, or obtain the information from a lighting manufacturer's catalog.

Commonly, as a rule of thumb, the following values are used for calculating **LLF or LLD:**

> **Fluorescent lamp 0.9**
>
> **Mercury lamp 0.83**
>
> **Metal-halide lamp 0.86**
>
> **High-pressure sodium lamp 0.94**
>
> **Tungsten-halogen lamp 0.96**
>
> **Incandescent lamp 0.95**

**L-73    LUMINAIRE DIRT DEPRECIATION (LDD)**

The reflectance, **diffusers,** and **lamps** within a fixture become dusty and dirty during the operation.

This will affect the lumen output of the fixture, and is called *luminaire dirt depreciation, or LDD.*

To determine the value of LLD, the following steps are given.

**Step 1.**    Please see Fig. L-37 under **"Maint. Cat."** column.

For each fixture a Roman numeral is given; for example, for fixture number 30, **"IV"** is given.

**Step 2.**    Now look at Fig. L-40 under **"category IV"** chart to determine the value of **LDD.**

**Note:**    a. Months indicates approximate cleaning period of fixtures.

b. The curves shown represent the condition of room from very clean to very dirty.

**298**

# Figure L-37 Coefficients of Utilization

Coefficients of Utilization for 20 Per Cent Effective Floor Cavity Reflectance ($\rho_{FC}$ = 20)

## 25 — Porcelain-enameled reflector with 35°CW shielding
Maint. Cat. II · SC 1.3 · 22½% ↑ / 65% ↓

| ρCC → | 80 | | | 70 | | | 50 | | | 30 | | | 10 | | | 0 | WDRC | RCR |
|---|---|---|---|---|---|---|---|---|---|---|---|---|---|---|---|---|---|---|
| ρW → / RCR ↓ | 50 | 30 | 10 | 50 | 30 | 10 | 50 | 30 | 10 | 50 | 30 | 10 | 50 | 30 | 10 | 0 | | ↓ |
| 0 | .99 | .99 | .99 | .94 | .94 | .94 | .85 | .85 | .85 | .77 | .77 | .77 | .69 | .69 | .69 | .65 | | 0 |
| 1 | .87 | .84 | .81 | .83 | .80 | .77 | .75 | .73 | .71 | .68 | .66 | .65 | .62 | .60 | .59 | .56 | .236 | 1 |
| 2 | .77 | .71 | .67 | .73 | .68 | .64 | .67 | .63 | .60 | .60 | .58 | .55 | .55 | .53 | .51 | .48 | .220 | 2 |
| 3 | .68 | .62 | .56 | .65 | .59 | .54 | .59 | .55 | .51 | .54 | .50 | .47 | .49 | .46 | .44 | .41 | .203 | 3 |
| 4 | .61 | .54 | .48 | .58 | .52 | .47 | .53 | .48 | .44 | .48 | .44 | .41 | .44 | .41 | .38 | .35 | .186 | 4 |
| 5 | .54 | .47 | .42 | .52 | .46 | .41 | .48 | .42 | .38 | .44 | .39 | .36 | .40 | .36 | .33 | .31 | .170 | 5 |
| 6 | .49 | .42 | .37 | .47 | .40 | .36 | .43 | .38 | .34 | .40 | .35 | .32 | .36 | .33 | .30 | .27 | .157 | 6 |
| 7 | .45 | .37 | .32 | .43 | .36 | .32 | .39 | .34 | .30 | .36 | .32 | .28 | .33 | .29 | .26 | .24 | .145 | 7 |
| 8 | .41 | .34 | .29 | .39 | .33 | .28 | .36 | .31 | .27 | .33 | .29 | .25 | .31 | .27 | .24 | .22 | .135 | 8 |
| 9 | .37 | .31 | .26 | .36 | .30 | .25 | .33 | .28 | .24 | .31 | .26 | .23 | .28 | .24 | .22 | .20 | .126 | 9 |
| 10 | .34 | .28 | .24 | .33 | .27 | .23 | .31 | .25 | .22 | .28 | .24 | .21 | .26 | .22 | .20 | .18 | .118 | 10 |

## 26 — Diffuse aluminum reflector with 35°CW shielding
Maint. Cat. II · SC 1.5/1.3 · 17% ↑ / 66% ↓

| ρCC → | 80 | | | 70 | | | 50 | | | 30 | | | 10 | | | 0 | WDRC | RCR |
|---|---|---|---|---|---|---|---|---|---|---|---|---|---|---|---|---|---|---|
| ρW → / RCR ↓ | 50 | 30 | 10 | 50 | 30 | 10 | 50 | 30 | 10 | 50 | 30 | 10 | 50 | 30 | 10 | 0 | | ↓ |
| 0 | .95 | .95 | .95 | .91 | .91 | .91 | .83 | .83 | .83 | .76 | .76 | .76 | .69 | .69 | .69 | .66 | | 0 |
| 1 | .85 | .82 | .79 | .81 | .79 | .76 | .75 | .73 | .71 | .69 | .67 | .66 | .63 | .62 | .61 | .58 | .197 | 1 |
| 2 | .75 | .71 | .67 | .72 | .68 | .65 | .67 | .63 | .61 | .62 | .59 | .57 | .57 | .55 | .53 | .51 | .194 | 2 |
| 3 | .67 | .61 | .57 | .65 | .59 | .55 | .60 | .56 | .52 | .55 | .52 | .49 | .51 | .49 | .46 | .44 | .184 | 3 |
| 4 | .60 | .54 | .49 | .58 | .52 | .48 | .54 | .49 | .45 | .50 | .46 | .43 | .46 | .43 | .41 | .39 | .173 | 4 |
| 5 | .54 | .47 | .43 | .52 | .46 | .42 | .49 | .43 | .40 | .45 | .41 | .38 | .42 | .39 | .36 | .34 | .162 | 5 |
| 6 | .49 | .42 | .37 | .47 | .41 | .37 | .44 | .39 | .35 | .41 | .37 | .33 | .38 | .35 | .32 | .30 | .151 | 6 |
| 7 | .44 | .38 | .33 | .43 | .37 | .32 | .40 | .35 | .31 | .38 | .33 | .30 | .35 | .31 | .28 | .27 | .141 | 7 |
| 8 | .40 | .34 | .29 | .39 | .33 | .29 | .37 | .31 | .28 | .34 | .30 | .27 | .32 | .28 | .26 | .24 | .132 | 8 |
| 9 | .37 | .31 | .26 | .36 | .30 | .26 | .34 | .29 | .25 | .32 | .27 | .24 | .30 | .26 | .23 | .21 | .124 | 9 |
| 10 | .34 | .28 | .24 | .33 | .27 | .23 | .31 | .26 | .23 | .29 | .25 | .22 | .28 | .24 | .21 | .19 | .117 | 10 |

## 27 — Porcelain-enameled reflector with 30°CW × 30°LW shielding
Maint. Cat. II · SC 1.0 · 23½% ↑ / 57% ↓

| ρCC → | 80 | | | 70 | | | 50 | | | 30 | | | 10 | | | 0 | WDRC | RCR |
|---|---|---|---|---|---|---|---|---|---|---|---|---|---|---|---|---|---|---|
| ρW → / RCR ↓ | 50 | 30 | 10 | 50 | 30 | 10 | 50 | 30 | 10 | 50 | 30 | 10 | 50 | 30 | 10 | 0 | | ↓ |
| 0 | .91 | .91 | .91 | .86 | .86 | .86 | .77 | .77 | .77 | .68 | .68 | .68 | .61 | .61 | .61 | .57 | | 0 |
| 1 | .80 | .77 | .75 | .76 | .74 | .71 | .69 | .67 | .65 | .62 | .60 | .59 | .55 | .54 | .53 | .50 | .182 | 1 |
| 2 | .71 | .67 | .63 | .68 | .64 | .60 | .61 | .58 | .55 | .55 | .52 | .49 | .50 | .48 | .46 | .43 | .174 | 2 |
| 3 | .63 | .58 | .53 | .60 | .55 | .51 | .55 | .51 | .47 | .50 | .46 | .44 | .45 | .42 | .40 | .38 | 163 | 3 |
| 4 | .57 | .51 | .46 | .54 | .49 | .44 | .49 | .45 | .41 | .45 | .41 | .38 | .41 | .38 | .35 | .33 | .151 | 4 |
| 5 | .51 | .45 | .40 | .49 | .43 | .39 | .45 | .40 | .36 | .41 | .37 | .34 | .37 | .34 | .31 | .29 | .140 | 5 |
| 6 | .46 | .40 | .35 | .44 | .38 | .34 | .41 | .36 | .32 | .37 | .33 | .30 | .34 | .30 | .28 | .26 | .130 | 6 |
| 7 | .42 | .36 | .31 | .40 | .35 | .30 | .37 | .32 | .29 | .34 | .30 | .27 | .31 | .28 | .25 | .23 | .121 | 7 |
| 8 | .38 | .32 | .28 | .37 | .31 | .27 | .34 | .29 | .26 | .31 | .27 | .24 | .29 | .25 | .23 | .21 | .113 | 8 |
| 9 | .35 | .29 | .25 | .34 | .28 | .25 | .31 | .27 | .23 | .29 | .25 | .22 | .27 | .23 | .21 | .19 | .106 | 9 |
| 10 | .33 | .27 | .23 | .31 | .26 | .22 | .29 | .24 | .21 | .27 | .23 | .20 | .25 | .21 | .19 | .17 | .099 | 10 |

## 28 — Diffuse aluminum reflector with 35°CW × 35°LW shielding
Maint. Cat. II · SC 1.5/1.1 · 17% ↑ / 56½% ↓

| ρCC → | 80 | | | 70 | | | 50 | | | 30 | | | 10 | | | 0 | WDRC | RCR |
|---|---|---|---|---|---|---|---|---|---|---|---|---|---|---|---|---|---|---|
| ρW → / RCR ↓ | 50 | 30 | 10 | 50 | 30 | 10 | 50 | 30 | 10 | 50 | 30 | 10 | 50 | 30 | 10 | 0 | | ↓ |
| 0 | .83 | .83 | .83 | .79 | .79 | .79 | .72 | .72 | .72 | .65 | .65 | .65 | .59 | .59 | .59 | .56 | | 0 |
| 1 | .74 | .72 | .70 | .71 | .69 | .67 | .65 | .63 | .62 | .59 | .58 | .57 | .54 | .53 | .52 | .50 | .160 | 1 |
| 2 | .66 | .62 | .59 | .64 | .60 | .57 | .58 | .56 | .53 | .54 | .51 | .49 | .49 | .47 | .46 | .44 | .158 | 2 |
| 3 | 59 | .54 | .50 | .57 | .53 | .49 | .53 | .49 | .46 | .48 | .46 | .43 | .45 | .42 | .40 | .38 | .150 | 3 |
| 4 | 53 | .48 | .44 | .51 | .46 | .42 | .47 | .43 | .40 | .44 | .41 | .38 | .40 | .38 | .36 | .34 | .141 | 4 |
| 5 | .48 | .42 | .38 | .46 | .41 | .37 | .43 | .39 | .35 | .40 | .37 | .34 | .37 | .34 | .32 | .30 | .132 | 5 |
| 6 | .44 | .38 | .34 | .42 | .37 | .33 | .39 | .35 | .31 | .36 | .33 | .30 | .34 | .31 | .28 | .27 | .124 | 6 |
| 7 | .40 | .34 | .30 | .38 | .33 | .29 | .36 | .31 | .28 | .33 | .30 | .27 | .31 | .28 | .25 | .24 | .116 | 7 |
| 8 | .36 | .31 | .27 | .35 | .30 | .26 | .33 | .28 | .25 | .31 | .27 | .24 | .29 | .25 | .23 | .21 | .109 | 8 |
| 9 | .33 | .28 | .24 | .32 | .27 | .24 | .30 | .26 | .23 | .28 | .24 | .22 | .26 | .23 | .21 | .19 | .102 | 9 |
| 10 | .31 | .25 | .22 | .30 | .25 | .22 | .28 | .24 | .21 | .26 | .22 | .20 | .25 | .21 | .19 | .18 | .096 | 10 |

## 29 — Metal or dense diffusing sides with 45°CW × 45°LW shielding
Maint. Cat. II · SC 1.1 · 39% ↑ / 32% ↓

| ρCC → | 80 | | | 70 | | | 50 | | | 30 | | | 10 | | | 0 | WDRC | RCR |
|---|---|---|---|---|---|---|---|---|---|---|---|---|---|---|---|---|---|---|
| ρW → / RCR ↓ | 50 | 30 | 10 | 50 | 30 | 10 | 50 | 30 | 10 | 50 | 30 | 10 | 50 | 30 | 10 | 0 | | ↓ |
| 0 | .75 | .75 | .75 | .69 | .69 | .69 | .57 | .57 | .57 | .46 | .46 | .46 | .37 | .37 | .37 | .32 | | 0 |
| 1 | .66 | .64 | .62 | .61 | .59 | .57 | .51 | .50 | .48 | .42 | .41 | .40 | .33 | .33 | .32 | .28 | .094 | 1 |
| 2 | .59 | .55 | .52 | .54 | .51 | .48 | .46 | .43 | .41 | .38 | .36 | .34 | .30 | .29 | .28 | .25 | .091 | 2 |
| 3 | .52 | .48 | .44 | .48 | .44 | .41 | .41 | .38 | .35 | .34 | .32 | .30 | .27 | .26 | .25 | .22 | .085 | 3 |
| 4 | .47 | .42 | .38 | .43 | .39 | .35 | .37 | .33 | .31 | .31 | .28 | .26 | .25 | .23 | .22 | .19 | .079 | 4 |
| 5 | .42 | .37 | .33 | .39 | .34 | .31 | .33 | .30 | .27 | .28 | .25 | .23 | .23 | .21 | .20 | .17 | .073 | 5 |
| 6 | .38 | .33 | .29 | .35 | .31 | .27 | .30 | .27 | .24 | .25 | .23 | .21 | .21 | .19 | .18 | .16 | .068 | 6 |
| 7 | .35 | .29 | .26 | .32 | .28 | .24 | .28 | .24 | .21 | .23 | .21 | .19 | .19 | .17 | .16 | .14 | .063 | 7 |
| 8 | .32 | .26 | .23 | .29 | .25 | .22 | .25 | .22 | .19 | .22 | .19 | .17 | .18 | .16 | .15 | .13 | .059 | 8 |
| 9 | .29 | .24 | .21 | .27 | .23 | .20 | .23 | .20 | .17 | .20 | .17 | .15 | .17 | .15 | .13 | .12 | .056 | 9 |
| 10 | .27 | .22 | .19 | .25 | .21 | .18 | .22 | .18 | .16 | .19 | .16 | .14 | .16 | .14 | .12 | .11 | .052 | 10 |

## 30 — Same as unit #29 except with top reflectors
Maint. Cat. IV · SC 1.0 · 6% ↑ / 46% ↓

| ρCC → | 80 | | | 70 | | | 50 | | | 30 | | | 10 | | | 0 | WDRC | RCR |
|---|---|---|---|---|---|---|---|---|---|---|---|---|---|---|---|---|---|---|
| ρW → / RCR ↓ | 50 | 30 | 10 | 50 | 30 | 10 | 50 | 30 | 10 | 50 | 30 | 10 | 50 | 30 | 10 | 0 | | ↓ |
| 0 | .61 | .61 | .61 | .58 | .58 | .58 | .55 | .55 | .55 | .51 | .51 | .51 | .48 | .48 | .48 | .46 | | 0 |
| 1 | .54 | .52 | .50 | .52 | .50 | .49 | .49 | .47 | .46 | .46 | .45 | .43 | .43 | .42 | .41 | .40 | .159 | 1 |
| 2 | .48 | .45 | .42 | .46 | .44 | .41 | .44 | .41 | .39 | .41 | .39 | .38 | .39 | .37 | .36 | .34 | .145 | 2 |
| 3 | .43 | .39 | .36 | .42 | .38 | .35 | .39 | .36 | .34 | .37 | .35 | .33 | .35 | .33 | .31 | .30 | .132 | 3 |
| 4 | .39 | .35 | .32 | .38 | .34 | .31 | .36 | .32 | .30 | .34 | .32 | .29 | .32 | .30 | .28 | .27 | .121 | 4 |
| 5 | .35 | .31 | .28 | .34 | .30 | .27 | .32 | .29 | .27 | .31 | .28 | .26 | .29 | .27 | .25 | .24 | .111 | 5 |
| 6 | .32 | .28 | .25 | .31 | .27 | .25 | .30 | .26 | .24 | .28 | .25 | .23 | .27 | .25 | .23 | .22 | .102 | 6 |
| 7 | .29 | .25 | .22 | .29 | .25 | .22 | .27 | .24 | .22 | .26 | .23 | .21 | .25 | .23 | .21 | .20 | .095 | 7 |
| 8 | .27 | .23 | .20 | .27 | .23 | .20 | .25 | .22 | .20 | .24 | .21 | .19 | .23 | .21 | .19 | .18 | .088 | 8 |
| 9 | .25 | .21 | .19 | .25 | .21 | .18 | .24 | .20 | .18 | .23 | .20 | .18 | .22 | .19 | .17 | .16 | .083 | 9 |
| 10 | .23 | .20 | .17 | .23 | 19 | .17 | .22 | .19 | .17 | .21 | .18 | .16 | .20 | .18 | .16 | .15 | .077 | 10 |

**Figure L-37  Coefficients of Utilization**

# Figure L-37 Coefficients of Utilization

| Typical Luminaire | | | Typical Intensity Distribution and Per Cent Lamp Lumens | ρCC → | 80 | | | 70 | | | 50 | | | 30 | | | 10 | | | 0 | WDRC | ρCC → RCR ↓ |
|---|---|---|---|---|---|---|---|---|---|---|---|---|---|---|---|---|---|---|---|---|---|---|
| | Maint. Cat. | SC | | ρW → RCR ↓ | 50 | 30 | 10 | 50 | 30 | 10 | 50 | 30 | 10 | 50 | 30 | 10 | 50 | 30 | 10 | 0 | | |
| | | | | | Coefficients of Utilization for 20 Per Cent Effective Floor Cavity Reflectance (ρFC = 20) | | | | | | | | | | | | | | | | | |
| 37 — 2-lamp diffuse wraparound—see note 7 | V | 1.3 | 8% / 37½% | 0 | .52 | .52 | .52 | .50 | .50 | .50 | .46 | .46 | .46 | .43 | .43 | .43 | .39 | .39 | .39 | .38 | | 0 |
| | | | | 1 | .44 | .42 | .40 | .42 | .40 | .39 | .39 | .37 | .36 | .36 | .35 | .33 | .33 | .32 | .31 | .30 | .201 | 1 |
| | | | | 2 | .38 | .35 | .32 | .37 | .33 | .31 | .34 | .31 | .29 | .31 | .29 | .27 | .28 | .27 | .25 | .24 | .171 | 2 |
| | | | | 3 | .33 | .29 | .26 | .32 | .28 | .25 | .29 | .26 | .24 | .27 | .25 | .22 | .25 | .23 | .21 | .20 | .149 | 3 |
| | | | | 4 | .29 | .25 | .22 | .28 | .24 | .21 | .26 | .23 | .20 | .24 | .21 | .19 | .22 | .20 | .18 | .17 | .132 | 4 |
| | | | | 5 | .26 | .22 | .19 | .25 | .21 | .18 | .23 | .20 | .17 | .21 | .18 | .16 | .20 | .17 | .15 | .14 | .117 | 5 |
| | | | | 6 | .23 | .19 | .16 | .22 | .18 | .16 | .21 | .17 | .15 | .19 | .16 | .14 | .18 | .15 | .13 | .12 | .106 | 6 |
| | | | | 7 | .21 | .17 | .14 | .20 | .16 | .14 | .19 | .15 | .13 | .17 | .15 | .12 | .16 | .14 | .12 | .11 | .096 | 7 |
| | | | | 8 | .19 | .15 | .12 | .18 | .15 | .12 | .17 | .14 | .12 | .16 | .13 | .11 | .15 | .12 | .11 | .10 | .088 | 8 |
| | | | | 9 | .17 | .14 | .11 | .17 | .13 | .11 | .16 | .13 | .10 | .15 | .12 | .10 | .14 | .11 | .09 | .09 | .081 | 9 |
| | | | | 10 | .16 | .12 | .10 | .15 | .12 | .10 | .14 | .11 | .09 | .14 | .11 | .09 | .13 | .10 | .09 | .08 | .075 | 10 |
| 38 — 4-lamp, 610 mm (2') wide troffer with 45° plastic louver—see note 7 | IV | 1.0 | 0% / 50% | 0 | .60 | .60 | .60 | .58 | .58 | .58 | .56 | .56 | .56 | .53 | .53 | .53 | .51 | .51 | .51 | .50 | | 0 |
| | | | | 1 | .53 | .51 | .49 | .52 | .50 | .49 | .50 | .48 | .47 | .48 | .47 | .46 | .46 | .45 | .44 | .43 | .168 | 1 |
| | | | | 2 | .47 | .44 | .42 | .46 | .43 | .41 | .44 | .42 | .40 | .43 | .41 | .39 | .41 | .40 | .38 | .37 | .159 | 2 |
| | | | | 3 | .42 | .38 | .36 | .41 | .38 | .35 | .40 | .37 | .35 | .39 | .36 | .34 | .37 | .35 | .34 | .32 | .146 | 3 |
| | | | | 4 | .38 | .34 | .31 | .37 | .34 | .31 | .36 | .33 | .30 | .35 | .32 | .30 | .34 | .32 | .30 | .29 | .135 | 4 |
| | | | | 5 | .34 | .30 | .27 | .34 | .30 | .27 | .33 | .29 | .27 | .32 | .29 | .27 | .31 | .28 | .26 | .25 | .124 | 5 |
| | | | | 6 | .31 | .27 | .24 | .31 | .27 | .24 | .30 | .27 | .24 | .29 | .26 | .24 | .28 | .26 | .24 | .23 | .114 | 6 |
| | | | | 7 | .29 | .25 | .22 | .28 | .24 | .22 | .28 | .24 | .22 | .27 | .24 | .21 | .26 | .23 | .21 | .20 | .106 | 7 |
| | | | | 8 | .26 | .22 | .20 | .26 | .22 | .20 | .25 | .22 | .20 | .25 | .22 | .20 | .24 | .21 | .19 | .19 | .099 | 8 |
| | | | | 9 | .24 | .21 | .18 | .24 | .21 | .18 | .24 | .20 | .18 | .23 | .20 | .18 | .23 | .20 | .18 | .17 | .092 | 9 |
| | | | | 10 | .23 | .19 | .17 | .22 | .19 | .17 | .22 | .19 | .16 | .22 | .19 | .16 | .21 | .18 | .16 | .16 | .086 | 10 |
| 39 — 4-lamp, 610 mm (2') wide troffer with 45° white metal louver—see note 7 | IV | 0.9 | 0% / 46% | 0 | .55 | .55 | .55 | .54 | .54 | .54 | .51 | .51 | .51 | .49 | .49 | .49 | .47 | .47 | .47 | .46 | | 0 |
| | | | | 1 | .49 | .48 | .46 | .48 | .47 | .46 | .46 | .45 | .44 | .44 | .43 | .43 | .43 | .42 | .42 | .41 | .137 | 1 |
| | | | | 2 | .44 | .42 | .40 | .43 | .41 | .39 | .42 | .40 | .38 | .40 | .39 | .37 | .39 | .38 | .37 | .36 | .131 | 2 |
| | | | | 3 | .40 | .37 | .34 | .39 | .36 | .34 | .38 | .36 | .33 | .37 | .35 | .33 | .36 | .34 | .32 | .32 | .122 | 3 |
| | | | | 4 | .36 | .33 | .30 | .36 | .33 | .30 | .35 | .32 | .30 | .34 | .31 | .29 | .33 | .31 | .29 | .28 | .113 | 4 |
| | | | | 5 | .33 | .30 | .27 | .33 | .29 | .27 | .32 | .29 | .27 | .31 | .28 | .26 | .30 | .28 | .26 | .25 | .104 | 5 |
| | | | | 6 | .30 | .27 | .24 | .30 | .27 | .24 | .29 | .26 | .24 | .29 | .26 | .24 | .28 | .25 | .24 | .23 | .097 | 6 |
| | | | | 7 | .28 | .25 | .22 | .28 | .24 | .22 | .27 | .24 | .22 | .27 | .24 | .22 | .26 | .23 | .22 | .21 | .090 | 7 |
| | | | | 8 | .26 | .23 | .20 | .26 | .22 | .20 | .25 | .22 | .20 | .25 | .22 | .20 | .24 | .22 | .20 | .19 | .085 | 8 |
| | | | | 9 | .24 | .21 | .19 | .24 | .21 | .19 | .23 | .20 | .18 | .23 | .20 | .18 | .23 | .20 | .18 | .18 | .079 | 9 |
| | | | | 10 | .23 | .19 | .17 | .22 | .19 | .17 | .22 | .19 | .17 | .22 | .19 | .17 | .21 | .19 | .17 | .16 | .075 | 10 |
| 40 — Fluorescent unit dropped diffuser, 4-lamp 610 mm (2') wide—see note 7 | V | 1.2 | 1% / 60½% | 0 | .73 | .73 | .73 | .71 | .71 | .71 | .68 | .68 | .68 | .65 | .65 | .65 | .62 | .62 | .62 | .60 | | 0 |
| | | | | 1 | .63 | .60 | .58 | .62 | .59 | .57 | .59 | .57 | .55 | .56 | .55 | .53 | .54 | .53 | .51 | .50 | .259 | 1 |
| | | | | 2 | .55 | .51 | .47 | .54 | .50 | .46 | .51 | .48 | .45 | .49 | .46 | .44 | .47 | .45 | .43 | .42 | .236 | 2 |
| | | | | 3 | .48 | .43 | .39 | .47 | .42 | .39 | .45 | .41 | .38 | .43 | .40 | .37 | .42 | .39 | .36 | .35 | .212 | 3 |
| | | | | 4 | .43 | .37 | .33 | .42 | .37 | .33 | .40 | .36 | .32 | .39 | .35 | .32 | .37 | .34 | .31 | .30 | .191 | 4 |
| | | | | 5 | .38 | .33 | .29 | .37 | .32 | .28 | .36 | .31 | .28 | .35 | .31 | .28 | .33 | .30 | .27 | .26 | .173 | 5 |
| | | | | 6 | .34 | .29 | .25 | .34 | .29 | .25 | .33 | .28 | .24 | .31 | .27 | .24 | .30 | .27 | .24 | .23 | .158 | 6 |
| | | | | 7 | .31 | .26 | .22 | .31 | .26 | .22 | .30 | .25 | .22 | .29 | .25 | .21 | .28 | .24 | .21 | .20 | .144 | 7 |
| | | | | 8 | .28 | .23 | .20 | .28 | .23 | .20 | .27 | .23 | .19 | .26 | .22 | .19 | .25 | .22 | .19 | .18 | .133 | 8 |
| | | | | 9 | .26 | .21 | .18 | .26 | .21 | .18 | .25 | .21 | .17 | .24 | .20 | .17 | .24 | .20 | .17 | .16 | .123 | 9 |
| | | | | 10 | .24 | .19 | .16 | .24 | .19 | .16 | .23 | .19 | .16 | .22 | .19 | .16 | .22 | .18 | .16 | .15 | .115 | 10 |
| 41 — Fluorescent unit with flat bottom diffuser, 4-lamp 610 mm (2') wide—see note 7 | V | 1.2 | 0% / 57½% | 0 | .69 | .69 | .69 | .67 | .67 | .67 | .64 | .64 | .64 | .61 | .61 | .61 | .59 | .59 | .59 | .58 | | 0 |
| | | | | 1 | .60 | .58 | .55 | .59 | .57 | .55 | .56 | .55 | .53 | .54 | .53 | .51 | .52 | .51 | .50 | .49 | .227 | 1 |
| | | | | 2 | .52 | .49 | .45 | .51 | .48 | .45 | .49 | .46 | .44 | .47 | .45 | .43 | .46 | .44 | .42 | .40 | .214 | 2 |
| | | | | 3 | .46 | .41 | .38 | .45 | .41 | .37 | .43 | .40 | .37 | .42 | .39 | .36 | .40 | .38 | .35 | .34 | .196 | 3 |
| | | | | 4 | .41 | .36 | .32 | .40 | .35 | .32 | .39 | .34 | .31 | .37 | .34 | .31 | .36 | .33 | .30 | .29 | .178 | 4 |
| | | | | 5 | .36 | .31 | .28 | .36 | .31 | .27 | .35 | .30 | .27 | .33 | .30 | .27 | .32 | .29 | .26 | .25 | .162 | 5 |
| | | | | 6 | .33 | .28 | .24 | .32 | .27 | .24 | .31 | .27 | .24 | .30 | .26 | .23 | .29 | .26 | .23 | .22 | .148 | 6 |
| | | | | 7 | .30 | .25 | .21 | .29 | .25 | .21 | .28 | .24 | .21 | .28 | .24 | .21 | .27 | .23 | .21 | .20 | .136 | 7 |
| | | | | 8 | .27 | .22 | .19 | .27 | .22 | .19 | .26 | .22 | .19 | .25 | .21 | .19 | .25 | .21 | .19 | .17 | .126 | 8 |
| | | | | 9 | .25 | .20 | .17 | .25 | .20 | .17 | .24 | .20 | .17 | .23 | .20 | .17 | .23 | .19 | .17 | .16 | .116 | 9 |
| | | | | 10 | .23 | .18 | .15 | .23 | .18 | .15 | .22 | .18 | .15 | .22 | .18 | .15 | .21 | .18 | .15 | .14 | .108 | 10 |
| 42 — Fluorescent unit with flat prismatic lens, 4-lamp 610 mm (2') wide—see note 7 | V | 1.4/1.2 | 0 / 63% | 0 | .75 | .75 | .75 | .73 | .73 | .73 | .70 | .70 | .70 | .67 | .67 | .67 | .64 | .64 | .64 | .63 | | 0 |
| | | | | 1 | .67 | .64 | .62 | .65 | .63 | .61 | .63 | .61 | .59 | .60 | .59 | .58 | .58 | .57 | .56 | .55 | .208 | 1 |
| | | | | 2 | .59 | .56 | .52 | .58 | .55 | .52 | .56 | .53 | .51 | .54 | .52 | .49 | .52 | .50 | .48 | .47 | .199 | 2 |
| | | | | 3 | .53 | .48 | .45 | .52 | .48 | .44 | .50 | .46 | .43 | .48 | .45 | .43 | .47 | .44 | .42 | .41 | .186 | 3 |
| | | | | 4 | .47 | .42 | .38 | .46 | .42 | .38 | .45 | .41 | .38 | .44 | .40 | .37 | .42 | .39 | .37 | .35 | .172 | 4 |
| | | | | 5 | .43 | .37 | .34 | .42 | .37 | .33 | .41 | .36 | .33 | .39 | .36 | .33 | .38 | .35 | .32 | .31 | .160 | 5 |
| | | | | 6 | .39 | .33 | .30 | .38 | .33 | .29 | .37 | .32 | .29 | .36 | .32 | .29 | .35 | .31 | .29 | .27 | .148 | 6 |
| | | | | 7 | .35 | .30 | .26 | .35 | .30 | .26 | .34 | .29 | .26 | .33 | .29 | .26 | .32 | .28 | .26 | .24 | .138 | 7 |
| | | | | 8 | .32 | .27 | .24 | .32 | .27 | .23 | .31 | .26 | .23 | .30 | .26 | .23 | .29 | .26 | .23 | .22 | .128 | 8 |
| | | | | 9 | .30 | .25 | .21 | .29 | .24 | .21 | .28 | .24 | .21 | .28 | .24 | .21 | .27 | .24 | .21 | .20 | .120 | 9 |
| | | | | 10 | .27 | .22 | .19 | .27 | .22 | .19 | .26 | .22 | .19 | .26 | .22 | .19 | .25 | .22 | .19 | .18 | .113 | 10 |

| Typical Luminaire | Typical Intensity Distribution and Per Cent Lamp Lumens | | ρCC → | 80 | | | 70 | | | 50 | | | 30 | | | 10 | | | 0 | | ρCC → |
|---|---|---|---|---|---|---|---|---|---|---|---|---|---|---|---|---|---|---|---|---|---|
| | | | ρW → | 50 | 30 | 10 | 50 | 30 | 10 | 50 | 30 | 10 | 50 | 30 | 10 | 50 | 30 | 10 | 0 | WDRC | ρW → |
| | Maint. Cat. | SC | RCR ↓ | Coefficients of Utilization for 20 Per Cent Effective Floor Cavity Reflectance (ρFC = 20) | | | | | | | | | | | | | | | | | RCR ↓ |
| 49 | I | 1 4/1.2 | 0 | 1.13 | 1.13 | 1.13 | 1.09 | 1.09 | 1.09 | 1.01 | 1.01 | 1.01 | .94 | .94 | .94 | .88 | .88 | .88 | .85 | | |
| | | | 1 | .95 | .90 | .86 | .92 | .87 | .83 | .85 | .82 | .78 | .79 | .76 | .74 | .74 | .72 | .69 | .66 | .464 | 1 |
| | | | 2 | .82 | .74 | .68 | .79 | .72 | .66 | .73 | .68 | .63 | .68 | .64 | .60 | .63 | .60 | .56 | .53 | .394 | 2 |
| | | | 3 | .71 | .62 | .55 | .69 | .61 | .54 | .64 | .57 | .52 | .59 | .54 | .49 | .55 | .51 | .47 | .44 | .342 | 3 |
| | | | 4 | .62 | .53 | .46 | .60 | .52 | .45 | .56 | .49 | .43 | .52 | .46 | .41 | .49 | .44 | .40 | .37 | .300 | 4 |
| | | | 5 | .55 | .46 | .39 | .54 | .45 | .39 | .50 | .43 | .37 | .47 | .40 | .36 | .44 | .38 | .34 | .32 | .267 | 5 |
| 2-lamp fluorescent strip unit with 235° reflector fluorescent lamps | | | 6 | .50 | .41 | .34 | .48 | .40 | .33 | .45 | .38 | .32 | .42 | .36 | .31 | .39 | .34 | .30 | .27 | .240 | 6 |
| | | | 7 | .45 | .36 | .30 | .43 | .35 | .29 | .41 | .34 | .28 | .38 | .32 | .27 | .36 | .30 | .26 | .24 | .218 | 7 |
| | | | 8 | .41 | .32 | .26 | .40 | .32 | .26 | .37 | .30 | .25 | .35 | .29 | .24 | .33 | .27 | .23 | .21 | .199 | 8 |
| | | | 9 | .37 | .29 | .24 | .36 | .23 | .23 | .34 | .27 | .22 | .32 | .26 | .22 | .30 | .25 | .21 | .19 | .183 | 9 |
| | | | 10 | .34 | .26 | .21 | .33 | .26 | .21 | .32 | .25 | .20 | .30 | .24 | .20 | .28 | .23 | .19 | .17 | .170 | 10 |

| Typical Luminaires | ρCC → | 80 | | | 70 | | | 50 | | | 30 | | | 10 | | | 0 |
|---|---|---|---|---|---|---|---|---|---|---|---|---|---|---|---|---|---|
| | ρW → | 50 | 30 | 10 | 50 | 30 | 10 | 50 | 30 | 10 | 50 | 30 | 10 | 50 | 30 | 10 | 0 |
| | RCR ↓ | Coefficients of utilization for 20 Per Cent Effective Floor Cavity Reflectance, ρFC | | | | | | | | | | | | | | |
| 50 | 1 | .42 | .40 | .39 | .36 | .35 | .33 | .25 | .24 | .23 | Coves are not recommended for lighting areas having low reflectances. | | | | | | |
| | 2 | .37 | .34 | .32 | .32 | .29 | .27 | .22 | .20 | .19 | | | | | | | |
| | 3 | .32 | .29 | .26 | .28 | .25 | .23 | .19 | .17 | .16 | | | | | | | |
| | 4 | .29 | .25 | .22 | .25 | .22 | .19 | .17 | .15 | .13 | | | | | | | |
| | 5 | .25 | .21 | .18 | .22 | .19 | .16 | .15 | .13 | .11 | | | | | | | |
| | 6 | .23 | .19 | .16 | .20 | .16 | .14 | .14 | .12 | .10 | | | | | | | |
| | 7 | .20 | .17 | .14 | .17 | .14 | .12 | .12 | .10 | .08 | | | | | | | |
| | 8 | .18 | .15 | .12 | .16 | .13 | .10 | .11 | .09 | .08 | | | | | | | |
| Single row fluorescent lamp cove without reflector, mult. by 0.93 for 2 rows and by 0.85 for 3 rows. | 9 | .17 | .13 | .10 | .15 | .11 | .09 | .10 | .08 | .07 | | | | | | | |
| | 10 | .15 | .12 | .09 | .13 | .10 | .08 | .09 | .07 | .06 | | | | | | | |
| 51 ρCC from below ~65% | 1 | | | | .60 | .58 | .56 | .58 | .56 | .54 | | | | | | | |
| | 2 | | | | .53 | .49 | .45 | .51 | .47 | .43 | | | | | | | |
| | 3 | | | | .47 | .42 | .37 | .45 | .41 | .36 | | | | | | | |
| | 4 | | | | .41 | .36 | .32 | .39 | .35 | .31 | | | | | | | |
| | 5 | | | | .37 | .31 | .27 | .35 | .30 | .26 | | | | | | | |
| | 6 | | | | .33 | .27 | .23 | .31 | .26 | .23 | | | | | | | |
| Diffusing plastic or glass. 1) Ceiling efficiency ~60%; diffuser transmittance ~50%; diffuser reflectance ~40%. Cavity with minimum obstructions and painted with 80% reflectance paint—use ρc = 70. 2) For lower reflectance paint or obstructions—use ρc = 50. | 7 | | | | .29 | .24 | .20 | .28 | .23 | .20 | | | | | | | |
| | 8 | | | | .26 | .21 | .18 | .25 | .20 | .17 | | | | | | | |
| | 9 | | | | .23 | .19 | .15 | .23 | .18 | .15 | | | | | | | |
| | 10 | | | | .21 | .17 | .13 | .21 | .16 | .13 | | | | | | | |
| 52 ρCC from below ~60% | 1 | | | | .71 | .68 | .66 | .67 | .66 | .65 | .65 | .64 | .62 | | | | |
| | 2 | | | | .63 | .60 | .57 | .61 | .58 | .55 | .59 | .56 | .54 | | | | |
| | 3 | | | | .57 | .53 | .49 | .55 | .52 | .48 | .54 | .50 | .47 | | | | |
| | 4 | | | | .52 | .47 | .43 | .50 | .45 | .42 | .48 | .44 | .42 | | | | |
| | 5 | | | | .46 | .41 | .37 | .44 | .40 | .37 | .43 | .40 | .36 | | | | |
| | 6 | | | | .42 | .37 | .33 | .41 | .36 | .32 | .40 | .35 | .32 | | | | |
| Prismatic plastic or glass. 1) Ceiling efficiency ~67%; prismatic transmittance ~72%; prismatic reflectance ~18%. Cavity with minimum obstructions and painted with 80% reflectance paint—use ρc = 70. 2) For lower reflectance paint or obstructions—use ρc = 50. | 7 | | | | .38 | .32 | .29 | .37 | .31 | .28 | .36 | .31 | .28 | | | | |
| | 8 | | | | .34 | .28 | .25 | .33 | .28 | .25 | .32 | .28 | .25 | | | | |
| | 9 | | | | .30 | .25 | .22 | .30 | .25 | .21 | .29 | .25 | .21 | | | | |
| | 10 | | | | .27 | .23 | .19 | .27 | .22 | .19 | .26 | .22 | .19 | | | | |
| 53 ρCC from below ~45% | 1 | | | | | | | .51 | .49 | .48 | | | | .47 | .46 | .45 | |
| | 2 | | | | | | | .46 | .44 | .42 | | | | .43 | .42 | .40 | |
| | 3 | | | | | | | .42 | .39 | .37 | | | | .39 | .38 | .36 | |
| | 4 | | | | | | | .38 | .35 | .33 | | | | .36 | .34 | .32 | |
| | 5 | | | | | | | .35 | .32 | .29 | | | | .33 | .31 | .29 | |
| | 6 | | | | | | | .32 | .29 | .26 | | | | .30 | .28 | .26 | |
| Louvered ceiling. 1) Ceiling efficiency ~50%; 45° shielding opaque louvers of 80% reflectance. Cavity with minimum obstructions and painted with 80% reflectance paint—use ρc = 50. 2) For other conditions refer to Fig. 6–18. | 7 | | | | | | | .29 | .26 | .23 | | | | .28 | .25 | .23 | |
| | 8 | | | | | | | .27 | .23 | .21 | | | | .26 | .23 | .21 | |
| | 9 | | | | | | | .24 | .21 | .19 | | | | .24 | .21 | .19 | |
| | 10 | | | | | | | .22 | .19 | .17 | | | | .22 | .19 | .17 | |

## Figure L-37  Coefficients of Utilization

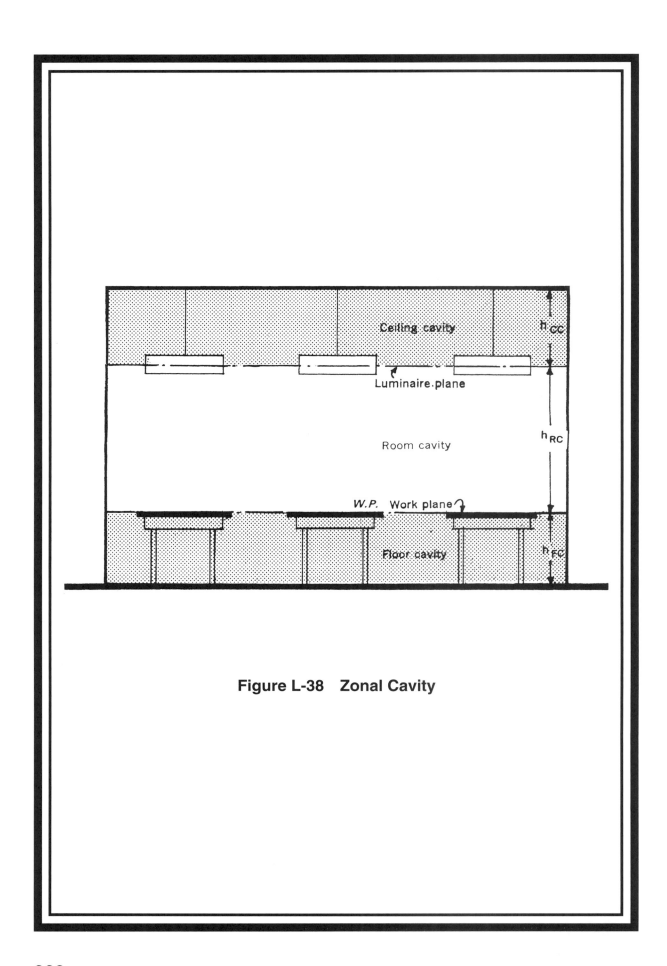

**Figure L-38   Zonal Cavity**

**Cavity Depth**

## Cavity Ratios

| Room Dimensions | | | | | | | | | | | | | | | | | |
|---|---|---|---|---|---|---|---|---|---|---|---|---|---|---|---|---|---|
| Width | Length | 1.0 | 1.5 | 2.0 | 2.5 | 3.0 | 3.5 | 4.0 | 5.0 | 6.0 | 7.0 | 8 | 9 | 10 | 11 | 12 |
| 8 | 8 | 1.2 | 1.9 | 2.5 | 3.1 | 3.7 | 4.4 | 5.0 | 6.2 | 7.5 | 8.8 | 10.0 | 11.2 | 12.5 | — | — |
| | 10 | 1.1 | 1.7 | 2.2 | 2.8 | 3.4 | 3.9 | 4.5 | 5.6 | 6.7 | 7.9 | 9.0 | 10.1 | 11.3 | 12.4 | — |
| | 14 | 1.0 | 1.5 | 2.0 | 2.5 | 3.0 | 3.4 | 3.9 | 4.9 | 5.9 | 6.9 | 7.8 | 8.8 | 9.7 | 10.7 | 11.7 |
| | 20 | 0.9 | 1.3 | 1.7 | 2.2 | 2.6 | 3.1 | 3.5 | 4.4 | 5.2 | 6.1 | 7.0 | 7.9 | 8.8 | 9.6 | 10.5 |
| | 30 | 0.8 | 1.2 | 1.6 | 2.0 | 2.4 | 2.8 | 3.2 | 4.0 | 4.7 | 5.5 | 6.3 | 7.1 | 7.9 | 8.7 | 9.5 |
| | 40 | 0.7 | 1.1 | 1.5 | 1.9 | 2.3 | 2.6 | 3.0 | 3.7 | 4.5 | 5.3 | 5.9 | 6.5 | 7.4 | 8.1 | 8.8 |
| 10 | 10 | 1.0 | 1.5 | 2.0 | 2.5 | 3.0 | 3.5 | 4.0 | 5.0 | 6.0 | 7.0 | 8.0 | 9.0 | 10.0 | 11.0 | 12.0 |
| | 14 | 0.9 | 1.3 | 1.7 | 2.1 | 2.6 | 3.0 | 3.4 | 4.3 | 5.1 | 6.0 | 6.9 | 7.8 | 8.6 | 9.5 | 10.4 |
| | 20 | 0.7 | 1.1 | 1.5 | 1.9 | 2.3 | 2.6 | 3.0 | 3.7 | 4.5 | 5.3 | 6.0 | 6.8 | 7.5 | 8.3 | 9.0 |
| | 30 | 0.7 | 1.0 | 1.3 | 1.7 | 2.0 | 2.3 | 2.7 | 3.3 | 4.0 | 4.7 | 5.3 | 6.0 | 6.6 | 7.3 | 8.0 |
| | 40 | 0.6 | 0.9 | 1.2 | 1.6 | 1.9 | 2.2 | 2.5 | 3.1 | 3.7 | 4.4 | 5.0 | 5.6 | 6.2 | 6.9 | 7.5 |
| | 60 | 0.6 | 0.9 | 1.2 | 1.5 | 1.7 | 2.0 | 2.3 | 2.9 | 3.5 | 4.1 | 4.7 | 5.3 | 5.9 | 6.5 | 7.1 |
| 12 | 12 | 0.8 | 1.2 | 1.7 | 2.1 | 2.5 | 2.9 | 3.3 | 4.2 | 5.0 | 5.8 | 6.7 | 7.5 | 8.4 | 9.2 | 10.0 |
| | 16 | 0.7 | 1.1 | 1.5 | 1.8 | 2.2 | 2.5 | 2.9 | 3.6 | 4.4 | 5.1 | 5.8 | 6.5 | 7.2 | 8.0 | 8.7 |
| | 24 | 0.6 | 0.9 | 1.3 | 1.6 | 1.9 | 2.2 | 2.5 | 3.1 | 3.7 | 4.4 | 5.0 | 5.6 | 6.2 | 6.9 | 7.5 |
| | 36 | 0.6 | 0.8 | 1.1 | 1.4 | 1.7 | 1.9 | 2.2 | 2.8 | 3.3 | 3.9 | 4.4 | 5.0 | 5.5 | 6.0 | 6.6 |
| | 50 | 0.5 | 0.8 | 1.0 | 1.3 | 1.5 | 1.8 | 2.1 | 2.6 | 3.1 | 3.6 | 4.1 | 4.6 | 5.1 | 5.6 | 6.2 |
| | 70 | 0.5 | 0.7 | 1.0 | 1.2 | 1.5 | 1.7 | 2.0 | 2.4 | 2.9 | 3.4 | 3.9 | 4.4 | 4.9 | 5.4 | 5.8 |
| 14 | 14 | 0.7 | 1.1 | 1.4 | 1.8 | 2.1 | 2.5 | 2.9 | 3.6 | 4.3 | 5.0 | 5.7 | 6.4 | 7.1 | 7.8 | 8.5 |
| | 20 | 0.6 | 0.9 | 1.2 | 1.5 | 1.8 | 2.1 | 2.4 | 3.0 | 3.6 | 4.2 | 4.9 | 5.5 | 6.1 | 6.7 | 7.3 |
| | 30 | 0.5 | 0.8 | 1.0 | 1.3 | 1.6 | 1.8 | 2.1 | 2.8 | 3.1 | 3.7 | 4.2 | 4.7 | 5.2 | 5.8 | 6.3 |
| | 42 | 0.5 | 0.7 | 1.0 | 1.2 | 1.4 | 1.7 | 1.9 | 2.4 | 2.9 | 3.3 | 3.8 | 4.3 | 4.7 | 5.2 | 5.7 |
| | 60 | 0.4 | 0.7 | 0.9 | 1.1 | 1.3 | 1.5 | 1.8 | 2.2 | 2.6 | 3.1 | 3.5 | 3.9 | 4.4 | 4.8 | 5.2 |
| | 90 | 0.4 | 0.6 | 0.8 | 1.0 | 1.2 | 1.4 | 1.6 | 2.0 | 2.5 | 2.9 | 3.3 | 3.7 | 4.1 | 4.5 | 5.0 |
| 17 | 17 | 0.6 | 0.9 | 1.2 | 1.5 | 1.8 | 2.1 | 2.3 | 2.9 | 3.5 | 4.1 | 4.7 | 5.3 | 5.9 | 6.5 | 7.0 |
| | 25 | 0.5 | 0.7 | 1.0 | 1.2 | 1.5 | 1.7 | 2.0 | 2.5 | 3.0 | 3.5 | 4.0 | 4.5 | 5.0 | 5.5 | 6.0 |
| | 35 | 0.4 | 0.7 | 0.9 | 1.1 | 1.3 | 1.5 | 1.7 | 2.2 | 2.6 | 3.1 | 3.5 | 3.9 | 4.4 | 4.8 | 5.2 |
| | 50 | 0.4 | 0.6 | 0.8 | 1.0 | 1.2 | 1.4 | 1.6 | 2.0 | 2.4 | 2.8 | 3.1 | 3.5 | 3.9 | 4.3 | 4.5 |
| | 80 | 0.4 | 0.5 | 0.7 | 0.9 | 1.1 | 1.2 | 1.4 | 1.8 | 2.1 | 2.5 | 2.9 | 3.3 | 3.6 | 4.0 | 4.3 |
| | 120 | 0.3 | 0.5 | 0.7 | 0.8 | 1.0 | 1.2 | 1.3 | 1.7 | 2.0 | 2.3 | 2.7 | 3.0 | 3.4 | 3.7 | 4.0 |
| 20 | 20 | 0.5 | 0.7 | 1.0 | 1.2 | 1.5 | 1.7 | 2.0 | 2.5 | 3.0 | 3.5 | 4.0 | 4.5 | 5.0 | 5.5 | 6.0 |
| | 30 | 0.4 | 0.6 | 0.8 | 1.0 | 1.2 | 1.5 | 1.7 | 2.1 | 2.5 | 2.9 | 3.3 | 3.7 | 4.1 | 4.5 | 4.9 |
| | 45 | 0.4 | 0.5 | 0.7 | 0.9 | 1.1 | 1.3 | 1.4 | 1.8 | 2.2 | 2.5 | 2.9 | 3.3 | 3.6 | 4.0 | 4.3 |
| | 60 | 0.3 | 0.5 | 0.7 | 0.8 | 1.0 | 1.2 | 1.3 | 1.7 | 2.0 | 2.3 | 2.7 | 3.0 | 3.4 | 3.7 | 4.0 |
| | 90 | 0.3 | 0.5 | 0.6 | 0.8 | 0.9 | 1.1 | 1.2 | 1.5 | 1.8 | 2.1 | 2.4 | 2.7 | 3.0 | 3.3 | 3.6 |
| | 150 | 0.3 | 0.4 | 0.6 | 0.7 | 0.8 | 1.0 | 1.1 | 1.4 | 1.7 | 2.0 | 2.3 | 2.6 | 2.9 | 3.2 | 3.4 |
| 24 | 24 | 0.4 | 0.6 | 0.8 | 1.0 | 1.2 | 1.5 | 1.7 | 2.1 | 2.5 | 2.9 | 3.3 | 3.7 | 4.1 | 4.5 | 5.0 |
| | 32 | 0.4 | 0.5 | 0.7 | 0.9 | 1.1 | 1.3 | 1.5 | 1.8 | 2.2 | 2.6 | 2.9 | 3.3 | 3.6 | 4.0 | 4.3 |
| | 50 | 0.3 | 0.5 | 0.6 | 0.8 | 0.9 | 1.2 | 1.2 | 1.5 | 1.8 | 2.2 | 2.5 | 2.8 | 3.1 | 3.4 | 3.7 |
| | 70 | 0.3 | 0.4 | 0.6 | 0.7 | 0.8 | 1.0 | 1.1 | 1.4 | 1.7 | 2.0 | 2.2 | 2.5 | 2.8 | 3.0 | 3.3 |
| | 100 | 0.3 | 0.4 | 0.5 | 0.6 | 0.8 | 0.9 | 1.0 | 1.3 | 1.6 | 1.8 | 2.1 | 2.4 | 2.6 | 2.9 | 3.1 |

**Figure L-39**

*IES Lighting Handbook 1984 Reference Volume*

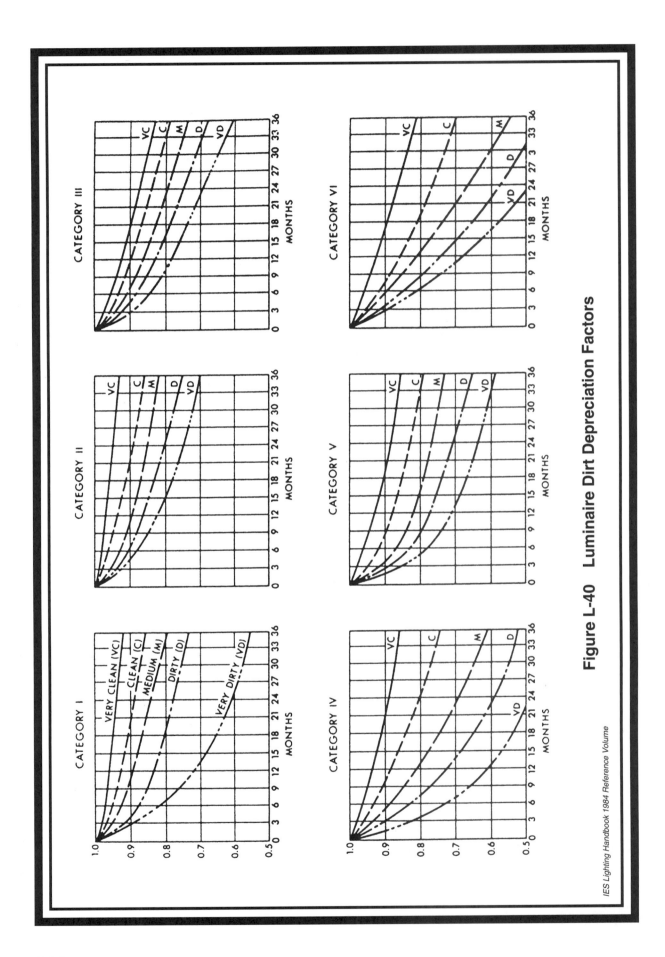

**Figure L-40 Luminaire Dirt Depreciation Factors**

IES Lighting Handbook 1984 Reference Volume

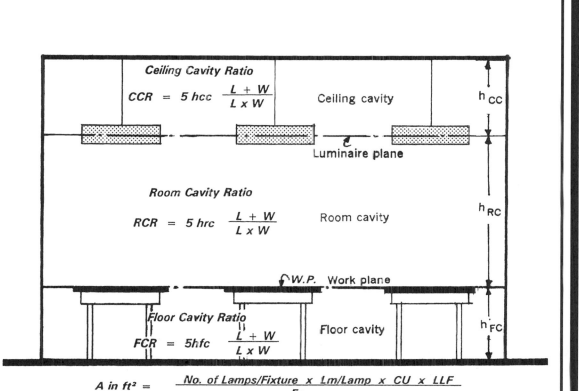

$$A \text{ in } ft^2 = \frac{\text{No. of Lamps/Fixture } \times \text{ Lm/Lamp } \times \text{ CU } \times \text{ LLF}}{Fc}$$

*A in ft² is the square feet each fixture illuminates.*

## Figure L-41  Zonal Cavity Calculation Method

## L-74 DESIGN AND CALCULATION CRITERIA FOR GENERAL LIGHTING

**Step 1.** Draw a cross section of the room or space indicating:

*a.* **Work-plane level**

*b.* **Location of fixtures**

Use Fig. L-41 as a guideline and show all dimensions required.

**Step 2. Find the footcandle** requirement for the room or space by using Fig. L-32.

**Step 3. Find the cavity ratio (CR).**

Use the calculation method given in L-70.

**Step 4. Choose the fixture type** by using the lighting manufacturer's catalog.

Fixture types are given in Fig. L-37 for reference only.

From the lighting catalog or Fig. L-37 obtain the size of the fixture, number of lamps, maintenance category (maint. cat.), and spacing requirements (sc).

Choose the type of lamp and obtain lm per lamp from Fig. L-28.

**Step 5. Determine the assumed reflectance** use the information given in L-68.

**Step 6. Find the actual reflectance** by using Fig. L-42*a* and *b*.

Use the assumed ceiling, wall, and floor reflectances found in step 5 and use the cavity ratio found in step 3. Use the **CCR** value to obtain the actual ceiling reflectance and the **FCR** value to obtain the actual floor reflectance.

**The assumed wall reflectance does not change by the movement of light, up or down.**

**Step 7. Find the coefficient of utilization (CU)** by using Fig. L-37, which is given for reference. These values are given in the lighting catalog for the fixture you have chosen.

See step 7 in L-75 for detailed information.

**Step 8.  Find light loss factor (LLF).**

Follow the guidelines given in L-71, L-72, and L-73.

**Step 9.  Find the total illumination in square feet that one fixture will provide on a work plane.**

$$\text{A in ft}^2 = \frac{\text{no. of lamps/fixture} \times \text{lm/lamp} \times \text{CU} \times \text{LLF}}{\text{fc}}$$

**A in ft²** is the square feet each fixture illuminates.

**Number of lamps/fixture** from step 4

**lm/lamp** from step 4

**CU** from step 7

**LLF** from step 8

**fc** from step 2

**Step 10.  Find out how many fixtures are needed in the room or space.**

$$\text{No. of fixtures} = \frac{\text{square feet each fixture illuminates}}{\text{total area of room or space}}$$

**Step 11.  Design the reflected ceiling plan.**

See step 11 in L-75 for detailed information.

*Example*

A classroom is 20 ft 6 in. wide, 30 ft 4 in. long, and has a floor-to-ceiling height of 12 ft 6 in.

The work plane is 2 ft 6 in. above the finished floor, and the center line of fixtures is 3 feet from the finished ceiling.

**Design a general lighting system for this classroom.**

*Solution*

**Step 1.** The cross section of the room has been drawn and is shown in Fig. L-41.

All dimensions have been converted to feet.

**Step 2. Find the footcandle requirements**

from Fig. L-32    **fc = 70**

**Step 3. Find the cavity ratio (CR).**

Using the calculation method from L-70:

$$\text{CCR} = 5 \times 3 \text{ ft} \times \frac{20.5 \text{ ft} + 30.3 \text{ ft}}{20.5 \text{ ft} \times 30.3 \text{ ft}} = 1.23 \approx 1.2$$

$$\text{RCR} = 5 \times 7 \text{ ft} \times \frac{20.5 \text{ ft} + 30.3 \text{ ft}}{20.5 \text{ ft} \times 30.3 \text{ ft}} = 2.87 \approx 2.8$$

$$\text{FCR} = 5 \times 2.5 \text{ ft} \times \frac{20.5 \text{ ft} + 30.3 \text{ ft}}{20.5 \text{ ft} + 30.3 \text{ ft}} = 1.025 \approx 1.0$$

Using L-69 and Fig. L-39 to check the **CR:**

**CCR** = 1.2

**RCR** = 2.9

**FCR** = 1

It is OK to use these CR values.

**Step 4. Choose a fixture type and lm/lamp**

*a.* From Fig. L-37 using fixture number 29, is a 4-ft fixture, with two lamps, Maint. Cat. is II and SC = 1.1

*b.* From Fig. L-28, using instant start lamp type T8 lm/lamp is 3725-3795. **Use 3725 lm/lamp.**

**Step 5. Determine the assumed reflectance** from L-68.

| | |
|---|---|
| Ceiling reflectance | 80% |
| Wall reflectance | 50% |
| Floor reflectance | 10% |

**Step 6. Find the actual reflectance**

*a.* Ceiling reflectance using Fig. L-42*b*

Using assumed reflectance 80%

Wall reflectance 50%, CCR = 1.2

**Actual ceiling reflectance = 64%**

*b.* **Wall reflectance never changes = 50%**

*c.* Floor reflectance using Fig. L-42*a*

Using assumed floor reflectance 10%

Wall reflectance 50%, FCR = 1

**Actual floor reflectance = 12%**

**Step 7. Find the coefficient of utilization (CU)** using Fig. L-37.

Lamp type number 29, RCR = 2.9

Actual ceiling reflectance = 64%

Wall reflectance = 50%

Using double interpolation to find actual CU:

70 ◄── 64 ◄── 50 ◄── ceiling reflectance

50 ◄─────────── 50 ◄── wall reflectance

| | | | | |
|---|---|---|---|---|
| **Table RCR** | 2 | 0.54 | | 0.46 |
| We have | 2.9 | 0.486 | 0.464 | 0.415 CU value |
| **Table RCR** | 3 | 0.48 | | 0.41 |

**Coefficient of utilization CU = 0.464**

**Step 8.  Find the light loss factor (LLF)**

     *a.*   From Fig. L-28 for lamp type T8 found in Step 4. Value for LLD is not given. Using LLD = 0.9. Please see L-72.

     *b.*   From Fig. L-40, using category II, medium clean, 16 months:

        LDD = 0.87

        **LLF = LLD $\times$ LDD**

        **LLF = 0.9 $\times$ 0.87 = 0.783**

**Step 9.  Find out how many square feet one fixture will illuminate (70 fc) on the work plane.**

$$A \text{ in ft}^2 = \frac{\text{no. of lamps/fixture} \times \text{lm/lamp} \times \text{CU} \times \text{LLF}}{\text{fc}}$$

$$A \text{ in ft}^2 = \frac{2 \text{ lamps} \times 3725 \text{ lm} \times 0.464 \times 0.783}{70} = 38.41 \text{ ft}^2$$

**Each fixture will illuminate 38.41 ft² of working plane with 70 fc.**

**Step 10.  Find out how many fixtures are needed in this classroom.**

     20.5 ft $\times$ 30.3 ft = 621.15 ft²

     621.15 ft² $\div$ 38.41 ft²/fix. = 16.17 fixtures

Use a minimum of 16 fixtures, maximum 18 fixtures.

**Step 11.** Design a reflected ceiling plan for this classroom and check the spacing requirements for the fixtures (sc).

Two reflected ceiling plans have been designed as A and B (Fig. L-43), and all dimensions are given.

From Fig. L-37, the required spacing for fixture number 29 under SC is 1.1. Fixture number 29 is 4 ft × 2 ft and the ceiling acoustical tiles to be used are 4 ft × 2 ft.

**Design A**

$$\frac{S}{M} = \frac{7 \text{ ft}}{7 \text{ ft}} = 1$$

where   S = spacing of fixtures (Fig. L-43), M = hrc (Fig. L-41)

SC is 1.1, about 10 percent overdesigned.

However, the spacing of fixtures is 7 ft 0 in. o.c. and does not allow the three ceiling tiles to be placed between the rows of fixtures which is not a proper design.

We are using 18 fixtures instead of 16, which is OK.

**Design B**

$$\frac{S}{M} = \frac{8 \text{ ft}}{7 \text{ ft}} = 1.14$$

SC is 1.1, about 4 percent underdesigned which is ok.

Because the spaces between the rows of fixtures are 8 ft 0 in. o.c., it does allow three ceiling tiles to be placed between the rows of fixtures which is a good design.

In both cases, design A and design B, the amount of illumination is acceptable; however, in design A the reflected ceiling plan is not preferred.

**Therefore, *use design B reflected ceiling plan.***

# Percent Effective Ceiling or Floor Cavity Reflectance

Percent Ceiling or Floor Reflectance

Percent Wall Reflectance

| Ceiling or Floor Cavity Rates | \ Ceiling 0 — Wall: 90 | 80 | 70 | 60 | 50 | 40 | 30 | 20 | 10 | 0 | Ceiling 10 — 90 | 80 | 70 | 60 | 50 | 40 | 30 | 20 | 10 | 0 | Ceiling 20 — 90 | 80 | 70 | 60 | 50 | 40 | 30 | 20 | 10 | 0 | Ceiling 30 — 90 | 80 | 70 | 60 | 50 | 40 | 30 | 20 | 10 | 0 | Ceiling 40 — 90 | 80 | 70 | 60 | 50 | 40 | 30 | 20 | 10 | 0 |
|---|---|---|---|---|---|---|---|---|---|---|---|---|---|---|---|---|---|---|---|---|---|---|---|---|---|---|---|---|---|---|---|---|---|---|---|---|---|---|---|---|---|---|---|---|---|---|---|---|---|---|
| 0.2 | 00 | 00 | 00 | 00 | 00 | 00 | 00 | 00 | 00 | 00 | 11 | 11 | 11 | 11 | 10 | 10 | 10 | 10 | 09 | 09 | 21 | 20 | 20 | 20 | 19 | 19 | 19 | 19 | 17 | 17 | 31 | 30 | 30 | 30 | 29 | 29 | 29 | 28 | 28 | 27 | 40 | 40 | 40 | 39 | 39 | 39 | 38 | 38 | 37 | 36 |

*Table continues across cavity rates 0.2 – 10.0 (rows: 0.2, 0.4, 0.6, 0.8, 1.0, 1.2, 1.4, 1.6, 1.8, 2.0, 2.2, 2.4, 2.6, 2.8, 3.0, 3.2, 3.4, 3.6, 3.8, 4.0, 4.2, 4.4, 4.6, 4.8, 5.0, 6.0, 7.0, 8.0, 9.0, 10.0) for each ceiling reflectance group (0, 10, 20, 30, 40) and wall reflectance value (90, 80, 70, 60, 50, 40, 30, 20, 10, 0).*

**Figure L-42a**

IES Lighting Handbook 1984 Reference Volume

# Percent Effective Ceiling or Floor Cavity Reflectance

Percent Ceiling or Floor Reflectance — Percent Wall Reflectance

**Percent Ceiling or Floor Reflectance = 90**

| Ceiling or Floor Cavity Ratio | Wall 90 | 80 | 70 | 60 | 50 | 40 | 30 | 20 | 10 | 0 |
|---|---|---|---|---|---|---|---|---|---|---|
| 0.2 | 89 | 88 | 88 | 87 | 86 | 85 | 84 | 84 | 84 | 82 |
| 0.4 | 88 | 87 | 86 | 85 | 84 | 83 | 81 | 80 | 79 | 76 |
| 0.6 | 87 | 86 | 85 | 83 | 82 | 80 | 77 | 76 | 74 | 73 |
| 0.8 | 87 | 85 | 83 | 82 | 80 | 77 | 75 | 73 | 71 | 67 |
| 1.0 | 86 | 83 | 80 | 77 | 75 | 72 | 69 | 66 | 64 | 62 |
| 1.2 | 85 | 82 | 78 | 75 | 72 | 69 | 66 | 63 | 60 | 57 |
| 1.4 | 85 | 80 | 77 | 73 | 69 | 65 | 62 | 59 | 57 | 52 |
| 1.6 | 84 | 79 | 75 | 71 | 67 | 63 | 59 | 56 | 53 | 50 |
| 1.8 | 83 | 78 | 73 | 69 | 64 | 60 | 56 | 53 | 50 | 48 |
| 2.0 | 83 | 77 | 72 | 67 | 62 | 58 | 54 | 50 | 47 | 43 |
| 2.2 | 82 | 76 | 70 | 65 | 59 | 54 | 50 | 47 | 44 | 40 |
| 2.4 | 82 | 75 | 69 | 64 | 58 | 53 | 49 | 45 | 43 | 37 |
| 2.6 | 81 | 74 | 67 | 62 | 56 | 51 | 46 | 42 | 38 | 35 |
| 2.8 | 81 | 73 | 66 | 60 | 54 | 49 | 44 | 40 | 36 | 34 |
| 3.0 | 80 | 72 | 64 | 58 | 52 | 47 | 42 | 38 | 34 | 30 |
| 3.2 | 79 | 71 | 63 | 56 | 50 | 45 | 40 | 36 | 32 | 28 |
| 3.4 | 79 | 70 | 62 | 54 | 49 | 43 | 38 | 34 | 30 | 27 |
| 3.6 | 78 | 69 | 61 | 53 | 47 | 42 | 36 | 32 | 28 | 25 |
| 3.8 | 78 | 69 | 60 | 51 | 45 | 40 | 35 | 31 | 27 | 23 |
| 4.0 | 77 | 69 | 58 | 51 | 44 | 39 | 33 | 29 | 25 | 22 |
| 4.2 | 77 | 62 | 57 | 50 | 43 | 37 | 32 | 28 | 24 | 21 |
| 4.4 | 76 | 61 | 56 | 49 | 42 | 36 | 31 | 27 | 23 | 20 |
| 4.6 | 76 | 60 | 55 | 47 | 40 | 35 | 30 | 26 | 21 | 18 |
| 4.8 | 75 | 59 | 54 | 46 | 39 | 34 | 28 | 25 | 21 | 16 |
| 5.0 | 75 | 59 | 53 | 45 | 38 | 33 | 28 | 24 | 20 | 16 |
| 6.0 | 73 | 61 | 50 | 41 | 34 | 29 | 24 | 20 | 16 | 11 |
| 7.0 | 70 | 58 | 45 | 38 | 30 | 25 | 21 | 18 | 14 | 08 |
| 8.0 | 68 | 55 | 42 | 35 | 27 | 23 | 18 | 15 | 12 | 06 |
| 9.0 | 66 | 52 | 38 | 31 | 25 | 21 | 16 | 14 | 11 | 05 |
| 10.0 | 65 | 51 | 36 | 29 | 22 | 19 | 15 | 11 | 09 | 04 |

**Percent Ceiling or Floor Reflectance = 80**

| Ceiling or Floor Cavity Ratio | Wall 90 | 80 | 70 | 60 | 50 | 40 | 30 | 20 | 10 | 0 |
|---|---|---|---|---|---|---|---|---|---|---|
| 0.2 | 79 | 78 | 78 | 77 | 76 | 76 | 75 | 74 | 74 | 72 |
| 0.4 | 78 | 77 | 77 | 76 | 75 | 74 | 73 | 72 | 71 | 68 |
| 0.6 | 78 | 76 | 75 | 73 | 72 | 70 | 68 | 66 | 65 | 63 |
| 0.8 | 78 | 75 | 73 | 71 | 69 | 67 | 65 | 63 | 61 | 57 |
| 1.0 | 77 | 74 | 72 | 69 | 67 | 65 | 62 | 60 | 57 | 55 |
| 1.2 | 76 | 73 | 70 | 67 | 64 | 61 | 58 | 55 | 53 | 51 |
| 1.4 | 76 | 72 | 68 | 65 | 62 | 59 | 55 | 53 | 50 | 48 |
| 1.6 | 75 | 71 | 67 | 63 | 60 | 57 | 53 | 50 | 47 | 45 |
| 1.8 | 75 | 70 | 66 | 62 | 58 | 55 | 51 | 48 | 45 | 44 |
| 2.0 | 74 | 69 | 64 | 60 | 56 | 52 | 48 | 45 | 42 | 38 |
| 2.2 | 74 | 68 | 63 | 58 | 54 | 50 | 45 | 42 | 38 | 35 |
| 2.4 | 73 | 67 | 61 | 56 | 52 | 47 | 43 | 40 | 36 | 33 |
| 2.6 | 73 | 66 | 60 | 55 | 50 | 45 | 41 | 38 | 34 | 31 |
| 2.8 | 72 | 65 | 59 | 53 | 48 | 43 | 39 | 36 | 32 | 30 |
| 3.0 | 72 | 65 | 58 | 52 | 47 | 42 | 37 | 34 | 30 | 27 |
| 3.2 | 72 | 65 | 57 | 51 | 45 | 40 | 36 | 32 | 30 | 25 |
| 3.4 | 71 | 64 | 56 | 49 | 44 | 39 | 34 | 32 | 27 | 23 |
| 3.6 | 71 | 63 | 54 | 48 | 43 | 38 | 32 | 28 | 25 | 22 |
| 3.8 | 70 | 62 | 53 | 47 | 41 | 36 | 31 | 27 | 23 | 19 |
| 4.0 | 70 | 61 | 53 | 46 | 40 | 35 | 30 | 26 | 22 | 16 |
| 4.2 | 69 | 60 | 52 | 45 | 39 | 34 | 29 | 25 | 21 | 18 |
| 4.4 | 69 | 60 | 51 | 44 | 38 | 33 | 28 | 24 | 20 | 16 |
| 4.6 | 68 | 59 | 50 | 43 | 37 | 31 | 26 | 23 | 19 | 15 |
| 4.8 | 68 | 58 | 49 | 42 | 36 | 30 | 25 | 21 | 18 | 14 |
| 5.0 | 68 | 58 | 48 | 41 | 35 | 30 | 25 | 21 | 17 | 11 |
| 6.0 | 66 | 55 | 44 | 38 | 31 | 27 | 22 | 19 | 15 | 10 |
| 7.0 | 64 | 53 | 41 | 35 | 28 | 24 | 19 | 16 | 14 | 07 |
| 8.0 | 62 | 50 | 38 | 32 | 25 | 21 | 17 | 13 | 11 | 05 |
| 9.0 | 61 | 49 | 36 | 30 | 23 | 19 | 15 | 13 | 10 | 04 |
| 10.0 | 59 | 46 | 33 | 27 | 21 | 18 | 14 | 10 | 08 | 03 |

**Percent Ceiling or Floor Reflectance = 70**

| Ceiling or Floor Cavity Ratio | Wall 90 | 80 | 70 | 60 | 50 | 40 | 30 | 20 | 10 | 0 |
|---|---|---|---|---|---|---|---|---|---|---|
| 0.2 | 70 | 69 | 68 | 68 | 67 | 67 | 66 | 66 | 65 | 64 |
| 0.4 | 69 | 68 | 67 | 67 | 66 | 66 | 65 | 65 | 64 | 58 |
| 0.6 | 69 | 67 | 66 | 65 | 64 | 63 | 61 | 59 | 57 | 54 |
| 0.8 | 68 | 66 | 64 | 62 | 60 | 58 | 56 | 55 | 53 | 50 |
| 1.0 | 68 | 65 | 62 | 60 | 58 | 55 | 53 | 52 | 50 | 47 |
| 1.2 | 67 | 64 | 61 | 59 | 57 | 54 | 51 | 48 | 46 | 44 |
| 1.4 | 67 | 63 | 60 | 58 | 55 | 51 | 49 | 45 | 43 | 41 |
| 1.6 | 67 | 62 | 59 | 56 | 53 | 47 | 45 | 43 | 40 | 38 |
| 1.8 | 66 | 61 | 58 | 54 | 51 | 46 | 44 | 41 | 38 | 36 |
| 2.0 | 66 | 60 | 56 | 52 | 49 | 45 | 42 | 38 | 36 | 33 |
| 2.2 | 66 | 60 | 55 | 51 | 48 | 43 | 40 | 37 | 35 | 32 |
| 2.4 | 65 | 60 | 54 | 50 | 45 | 43 | 38 | 35 | 33 | 30 |
| 2.6 | 65 | 59 | 54 | 49 | 44 | 41 | 37 | 34 | 31 | 28 |
| 2.8 | 65 | 58 | 53 | 48 | 43 | 40 | 36 | 32 | 30 | 26 |
| 3.0 | 64 | 58 | 52 | 47 | 42 | 38 | 34 | 31 | 28 | 24 |
| 3.2 | 64 | 57 | 51 | 45 | 40 | 36 | 31 | 28 | 25 | 23 |
| 3.4 | 64 | 57 | 50 | 44 | 39 | 35 | 30 | 27 | 24 | 22 |
| 3.6 | 63 | 56 | 49 | 43 | 38 | 33 | 28 | 25 | 22 | 21 |
| 3.8 | 63 | 56 | 49 | 42 | 37 | 32 | 27 | 24 | 21 | 19 |
| 4.0 | 63 | 55 | 48 | 42 | 36 | 31 | 26 | 23 | 20 | 17 |
| 4.2 | 62 | 54 | 47 | 41 | 35 | 30 | 25 | 22 | 19 | 16 |
| 4.4 | 62 | 54 | 46 | 40 | 34 | 28 | 24 | 21 | 18 | 13 |
| 4.6 | 62 | 53 | 45 | 39 | 32 | 27 | 23 | 19 | 16 | 13 |
| 4.8 | 62 | 53 | 45 | 38 | 32 | 27 | 22 | 18 | 16 | 12 |
| 5.0 | 61 | 52 | 44 | 36 | 31 | 26 | 22 | 17 | 16 | 12 |
| 6.0 | 59 | 49 | 41 | 34 | 28 | 24 | 20 | 16 | 13 | 09 |
| 7.0 | 56 | 46 | 38 | 31 | 25 | 21 | 18 | 14 | 11 | 07 |
| 8.0 | 55 | 45 | 35 | 29 | 23 | 19 | 15 | 13 | 10 | 05 |
| 9.0 | 53 | 43 | 33 | 27 | 21 | 18 | 14 | 11 | 09 | 04 |
| 10.0 | 52 | 41 | 31 | 25 | 19 | 16 | 12 | 10 | 08 | 03 |

**Percent Ceiling or Floor Reflectance = 60**

| Ceiling or Floor Cavity Ratio | Wall 90 | 80 | 70 | 60 | 50 | 40 | 30 | 20 | 10 | 0 |
|---|---|---|---|---|---|---|---|---|---|---|
| 0.2 | 60 | 59 | 59 | 58 | 57 | 56 | 56 | 55 | 54 | 53 |
| 0.4 | 60 | 59 | 58 | 57 | 56 | 55 | 54 | 53 | 52 | 50 |
| 0.6 | 60 | 58 | 57 | 56 | 54 | 53 | 51 | 50 | 48 | 46 |
| 0.8 | 60 | 57 | 56 | 54 | 52 | 51 | 48 | 47 | 46 | 43 |
| 1.0 | 59 | 57 | 55 | 53 | 51 | 48 | 46 | 44 | 43 | 41 |
| 1.2 | 59 | 56 | 54 | 51 | 49 | 46 | 44 | 41 | 39 | 38 |
| 1.4 | 59 | 55 | 53 | 50 | 47 | 45 | 41 | 39 | 38 | 36 |
| 1.6 | 59 | 55 | 52 | 49 | 46 | 43 | 39 | 37 | 35 | 33 |
| 1.8 | 58 | 54 | 51 | 47 | 44 | 42 | 38 | 36 | 33 | 31 |
| 2.0 | 58 | 54 | 50 | 46 | 43 | 41 | 37 | 34 | 32 | 29 |
| 2.2 | 58 | 53 | 49 | 44 | 42 | 37 | 35 | 31 | 28 | 26 |
| 2.4 | 58 | 53 | 48 | 43 | 39 | 35 | 33 | 30 | 27 | 24 |
| 2.6 | 58 | 53 | 48 | 42 | 39 | 35 | 31 | 28 | 26 | 23 |
| 2.8 | 58 | 52 | 47 | 41 | 37 | 33 | 29 | 26 | 24 | 21 |
| 3.0 | 57 | 52 | 46 | 40 | 37 | 32 | 28 | 25 | 23 | 20 |
| 3.2 | 57 | 51 | 45 | 39 | 35 | 31 | 27 | 23 | 22 | 18 |
| 3.4 | 57 | 51 | 45 | 40 | 38 | 30 | 26 | 22 | 20 | 17 |
| 3.6 | 57 | 50 | 44 | 38 | 33 | 28 | 25 | 21 | 19 | 16 |
| 3.8 | 57 | 50 | 44 | 38 | 32 | 28 | 24 | 20 | 17 | 15 |
| 4.0 | 57 | 49 | 42 | 37 | 31 | 27 | 23 | 20 | 17 | 14 |
| 4.2 | 56 | 49 | 42 | 37 | 31 | 27 | 23 | 19 | 17 | 14 |
| 4.4 | 56 | 49 | 41 | 36 | 30 | 24 | 22 | 18 | 16 | 13 |
| 4.6 | 56 | 49 | 41 | 35 | 30 | 25 | 21 | 18 | 15 | 12 |
| 4.8 | 56 | 48 | 40 | 34 | 29 | 24 | 20 | 16 | 14 | 10 |
| 5.0 | 56 | 48 | 40 | 34 | 29 | 24 | 20 | 16 | 14 | 11 |
| 6.0 | 54 | 45 | 37 | 31 | 26 | 22 | 18 | 14 | 11 | 08 |
| 7.0 | 53 | 43 | 35 | 28 | 23 | 19 | 15 | 12 | 10 | 05 |
| 8.0 | 52 | 42 | 33 | 26 | 20 | 15 | 13 | 10 | 07 | 04 |
| 9.0 | 52 | 40 | 31 | 24 | 18 | 14 | 11 | 09 | 07 | 03 |
| 10.0 | 52 | 39 | 29 | 24 | 17 | 14 | 10 | 08 | 06 | 02 |

**Percent Ceiling or Floor Reflectance = 50**

| Ceiling or Floor Cavity Ratio | Wall 90 | 80 | 70 | 60 | 50 | 40 | 30 | 20 | 10 | 0 |
|---|---|---|---|---|---|---|---|---|---|---|
| 0.2 | 50 | 50 | 50 | 49 | 48 | 48 | 47 | 47 | 46 | 44 |
| 0.4 | 50 | 49 | 49 | 48 | 48 | 46 | 45 | 44 | 44 | 42 |
| 0.6 | 50 | 48 | 48 | 47 | 46 | 45 | 43 | 42 | 41 | 38 |
| 0.8 | 50 | 48 | 47 | 46 | 44 | 43 | 40 | 39 | 38 | 36 |
| 1.0 | 50 | 48 | 46 | 45 | 43 | 41 | 38 | 37 | 36 | 34 |
| 1.2 | 50 | 47 | 45 | 43 | 41 | 39 | 36 | 35 | 34 | 29 |
| 1.4 | 50 | 47 | 45 | 42 | 40 | 38 | 35 | 34 | 34 | 27 |
| 1.6 | 50 | 47 | 44 | 41 | 38 | 36 | 34 | 32 | 30 | 26 |
| 1.8 | 50 | 46 | 43 | 40 | 37 | 35 | 32 | 30 | 28 | 25 |
| 2.0 | 50 | 46 | 43 | 40 | 37 | 34 | 31 | 30 | 28 | 24 |
| 2.2 | 50 | 46 | 42 | 39 | 36 | 33 | 29 | 27 | 24 | 22 |
| 2.4 | 50 | 46 | 42 | 38 | 35 | 31 | 28 | 26 | 23 | 21 |
| 2.6 | 50 | 46 | 41 | 37 | 34 | 30 | 26 | 25 | 22 | 20 |
| 2.8 | 50 | 45 | 41 | 36 | 34 | 30 | 25 | 24 | 20 | 19 |
| 3.0 | 50 | 45 | 40 | 36 | 32 | 28 | 24 | 21 | 20 | 17 |
| 3.2 | 50 | 44 | 39 | 35 | 31 | 27 | 23 | 22 | 20 | 16 |
| 3.4 | 50 | 44 | 39 | 35 | 30 | 26 | 22 | 19 | 17 | 15 |
| 3.6 | 50 | 44 | 38 | 34 | 29 | 25 | 21 | 18 | 16 | 14 |
| 3.8 | 50 | 44 | 38 | 34 | 29 | 25 | 21 | 18 | 15 | 13 |
| 4.0 | 50 | 44 | 38 | 33 | 28 | 24 | 20 | 17 | 15 | 12 |
| 4.2 | 50 | 43 | 37 | 32 | 28 | 23 | 20 | 17 | 14 | 12 |
| 4.4 | 50 | 43 | 37 | 32 | 28 | 23 | 19 | 16 | 13 | 11 |
| 4.6 | 50 | 43 | 36 | 31 | 26 | 22 | 18 | 15 | 12 | 10 |
| 4.8 | 50 | 43 | 36 | 30 | 25 | 22 | 18 | 15 | 12 | 09 |
| 5.0 | 50 | 43 | 35 | 30 | 25 | 21 | 17 | 14 | 12 | 09 |
| 6.0 | 50 | 42 | 34 | 29 | 23 | 19 | 15 | 13 | 10 | 06 |
| 7.0 | 49 | 41 | 32 | 25 | 21 | 18 | 14 | 11 | 08 | 05 |
| 8.0 | 49 | 40 | 30 | 25 | 19 | 16 | 12 | 10 | 07 | 03 |
| 9.0 | 48 | 39 | 29 | 24 | 18 | 15 | 11 | 09 | 07 | 03 |
| 10.0 | 47 | 37 | 27 | 22 | 17 | 14 | 10 | 08 | 06 | 02 |

Ceiling or Floor Cavity Ratios

## Figure L-42b (continued)

IES Lighting Handbook 1984 Reference Volume

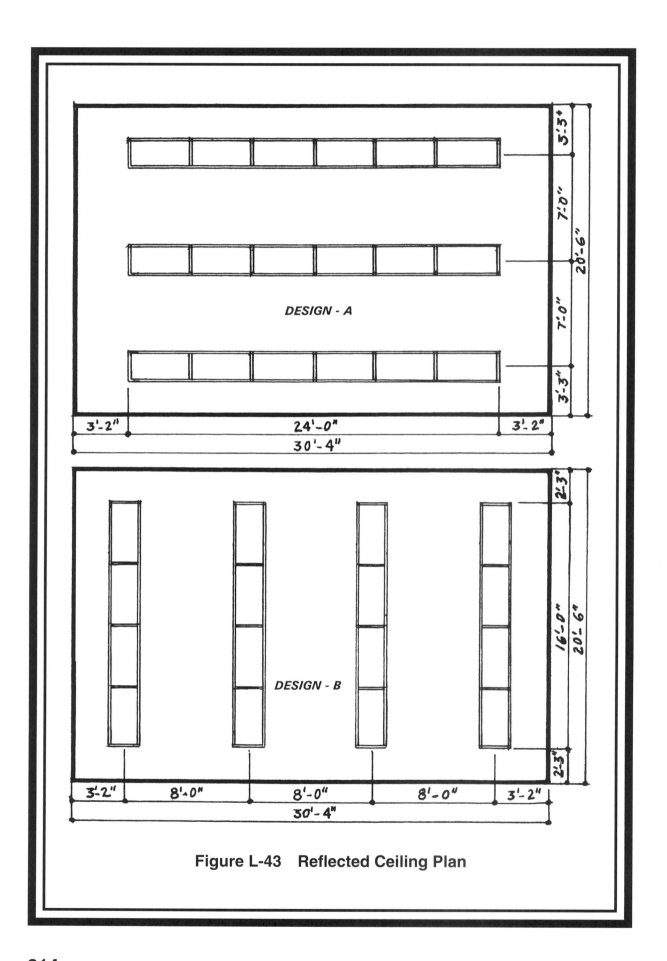

Figure L-43    Reflected Ceiling Plan

314

# CONTENTS

Part 8

# VERTICAL TRANSPORTATION SYSTEMS

## Vertical Transportation Systems

## Elevators

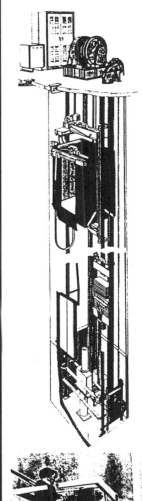

# Elevator Selection

# Vertical Transportation Systems for Handicapped Persons

# Escalators and Moving Ramps

# VERTICAL TRANSPORTATION SYSTEMS

## V-1    GENERAL INFORMATION

Every floor of every building which is designed to be used by the public **must** be accessible to physically handicapped persons. **"It is the law";** therefore, it is essential to provide at least one elevator in any public building of two stories or more. Please see V-26.

Furthermore, elevators have become an integral part of any commercial building of two or more stories, and the availability of elevator services plays a great role in a prestige building.

The vertical movement of people and goods from floor to floor in a building requires the evaluation and understanding of the following in order to produce an adequate quantity and quality of service and performance:

a.    **What is the nature and application of the vertical transportation?**

b.    **How does it function?**

c.    **What are the circulation and space requirements?**

Today, architectural practice **demands** the knowledge and understanding of how the **selection** of vertical transportation is designed and calculated, because

a.    We **cannot** begin to design a successful multistory building(s) without knowing the circulation and space requirements for vertical transportation.

b.    We **cannot** afford calling a consultant for help every time we get involved in designing a multistory building.

   1.    **It is costly.**

   2.    **It is time-consuming.**

   3.    **It is confusing.**

   4.    **It is not necessary.**

The design and calculation for selecting a vertical transportation system for all types of buildings, regardless of size:

a.    **Require a basic mathematical knowledge.**

b.    **Is simple to design and calculate.**

c.    **Consumes very little time.**

d.    **Is rewarding in both self-satisfaction and finances.**

**317**

**Figure V-1   Early "Man-Powered" Walking Hoist**

Many components which produce a building (masonry materials, doors, windows, lumber, steel, appliances, equipment, etc.) have been *pre-engineered* or *standard-designed* in many different sizes, shapes, or forms, and they are available to us for incorporating into our structures without any difficulty because we have learned:

a. **What they are.**

b. **How they have been designed.**

c. **How they have been constructed or made.**

d. **What their function is.**

e. **How they are built into or installed in a building.**

Vertical transportation systems are also **pre-engineered,** which is another way of saying that **standard-designed elevators, escalators, etc.,** have been engineered and manufactured in quantity to help reduce the costs. They are available in many types, capacities, and speeds.

The exact dimensions and specifications are available, which frees us from a lot of engineering and other detailed work.

In the following pages we will find answers to the following questions:

a. **What is vertical transportation?** And,

b. **How do we use it?**

In order to honor **Elisha Graves Otis,** the founder of the Otis Elevator Company, whose invention of the *safety elevator* in 1854 started the elevator industry, which has created both pleasure and convenience in our lives, some of the drawings, charts, tables, and information used in this part are based on data obtained from the Otis Elevator Company.

"Elevonic, Otis, and Escal-Aire are the trademarks of the Otis Elevator Co."

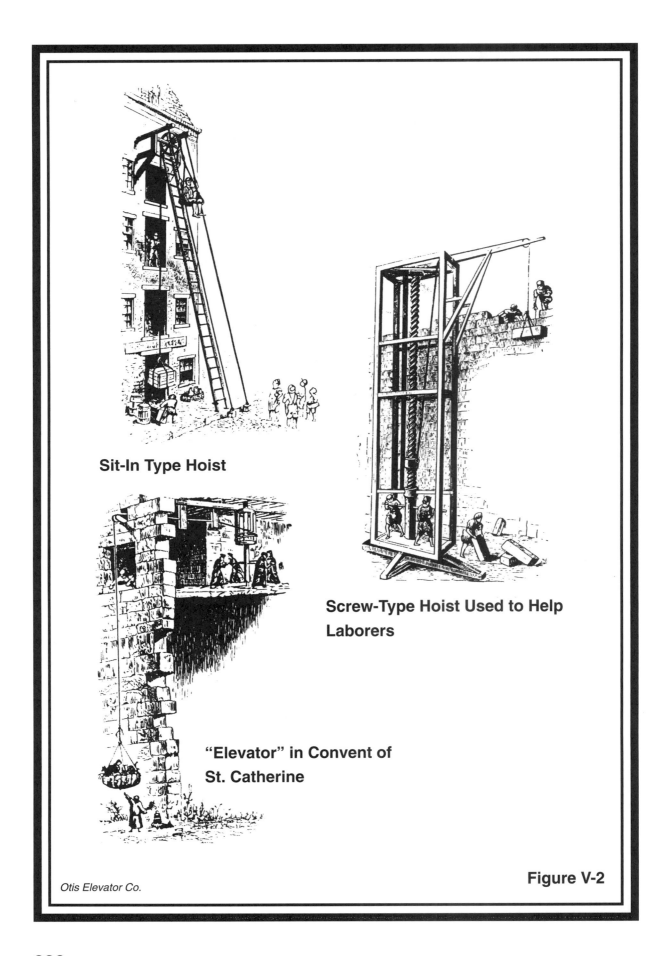

**Sit-In Type Hoist**

**Screw-Type Hoist Used to Help Laborers**

**"Elevator" in Convent of St. Catherine**

*Otis Elevator Co.*

**Figure V-2**

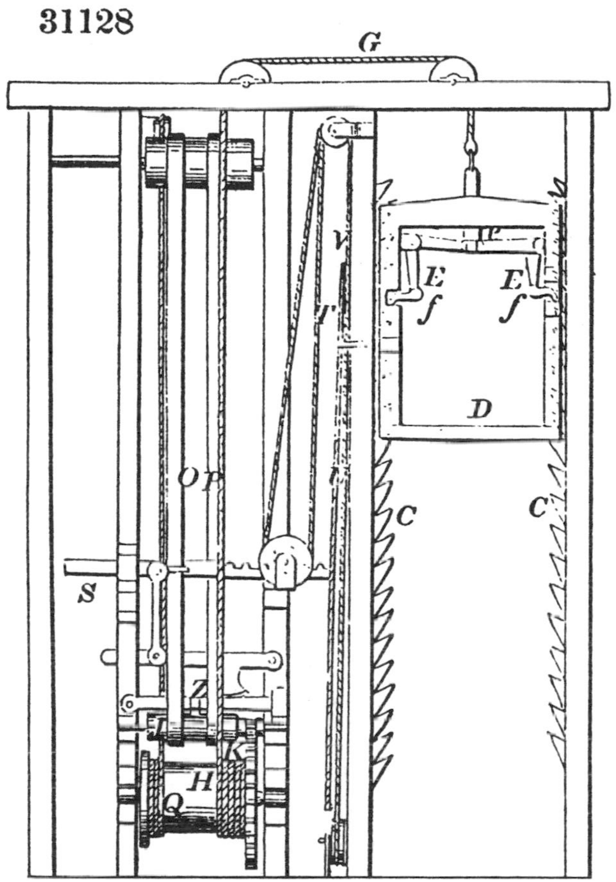

**Figure V-4** Elisha Otis giving a public demonstration of his safety elevator at the Crystal Palace Exposition, New York City, in 1854.

31128

**Figure V-3** Drawing from patent granted to Elisha Otis for elevator safety device. Patent also covered a drum counterweight.

*Otis Elevator Co.*

321

| | |
|---|---|
| **2680 B.C.** | The Egyptians built the Great Pyramid of Khufu at Gizeh near Cairo. It is 756 feet square and 482 feet in height, built with stone blocks weighing over 200 tons each. They must have employed some type of *hoists* to build this pyramid. |
| **236 B.C.** | **Archimedes,** a scientist, developed a **hoisting device** operated by ropes. |
| **200 B.C.** | A human-powered **walking hoist** was developed (Fig. V-1). |
| **80 A.D.** | A **crude elevator**(s) was used in the Roman Colosseum to raise the gladiators and wild animals to the level of the arena. |
| **500 A.D. to 1200 A.D.** | **Hoists** were used to raise people and supplies to isolated locations. The monastery of St. Barlaam in Greece was built 200 feet above the ground, with a **hoist** using either a cargo net or a basket as the only method of access. |
| **1203** | In France, a **hoist** was developed with a **tread-wheel**; the walking power used was a donkey. For many years, different types of hoists and lifts were developed for various uses (Fig. V-2). |
| **1835** | A steam-power-driven elevator called the *teagle* was used in England. |
| **1845** | **Sir William Thompson** developed the first *hydraulic elevator.* |
| **1850** | In the United States, **commercial elevators** operating with steam were used for freight service. |
| **1852** | **Elisha Graves Otis,** the founder of the Otis Elevator Company, invented the *safety brake* for elevators (Fig. V-3). |
| **1854** | E. G. Otis exhibited his complete safety elevator at the Crystal Palace Exposition in New York City. This was the start of the *elevator industry* (Fig. V-4). |
| **1857** | Otis installed the **first passenger elevator** at E. V. Haughwout and Company, a five-story building in New York City, with a 1000-pound capacity and a speed of 400 feet per minute (Fig. V-5). |

| 1878 | The **first Otis** *hydraulic* **passenger** elevator was installed in a 111-foot building at 155 Broadway in New York City. |
|---|---|
| 1889 | The **first DC** *elevator machines* were installed in the Demarest Building in New York City with a speed of 100 feet per minute. |
| 1903 | The **first** *gearless traction* **elevators** were installed in the 182-foot Beaver Building in New York City. The elevators had a 2500-pound capacity and operated at a speed of 500 feet per minute. |

## THE ERA OF TALL BUILDINGS BEGINS

| 1906 | 41-story Singer Building. |
|---|---|
| 1908 | 52-story Metropolitan Life Insurance Building. |
| 1912 | 60-story Woolworth Building. |
| 1915 | Otis developed the *microdrive,* a self-leveling device for elevators. |
| 1929 | 71-story Bank of Manhattan Building, New York City. |
| 1931 | 66-story Wall Tower Building, New York City. 77-story Chrysler Building, New York City. 102-story Empire State Building, New York City. |
| 1950 | The **first** *autotronic* **elevators** without attendants were installed in the Atlantic Refining Building in Dallas, Texas. |
| 1966 | 100-story John Hancock Center, Chicago. |
| 1970 to 1994 | The two tallest buildings in the world were constructed. |
| | 1. World Trade Center (the "twin towers") in New York City, 1350 feet in height. |
| | 2. The Sears Tower in Chicago, 1450 feet in height. |
| 1979 | The **first fully integrated** *microcomputer system,* called "Elevonic 101," designed and developed by the Otis Elevator Company to control every aspect of the elevator operation systems, was put into operation. |

**Figure V-5    E. V. Haughwout & Co. building at Broadway and Broome St., New York, where first passenger elevator was installed in 1857.**

324

## V-3    CODES AND REGULATIONS

There are several codes and regulations governing the installation and operation of vertical and horizontal means of transportation.

### American National Standards Institute (ANSI)

*ANSI Code A-17.1* **"Safety Code"** for elevators, escalators, dumbwaiters, and moving walks. This code has legal force in many cities in the United States.

Many states and cities within the United States have their own codes for vertical and horizontal transportation systems.

### The National Fire Protection Association (N.F.P.A.)

*NFPA No. 101.* **"Life Safety Code"** has binding requirements for the fire safety of all vertical and horizontal transportation systems.

*NFPA No. 70.* **"The National Electric Code"** has requirements for the electrical wiring and all electric devices used in vertical and horizontal transportation systems.

**American National Standards Institute** has specifications for use of elevators by physically handicapped persons, in **Sec. 5.9, "Elevators."** The **ANSI** Code has provisions for physically handicapped persons.

The **National Elevator Industry Incorporated** (NEII) has standards for elevator layouts and installations and standards for physically handicapped persons. Please see V-26.

We architects have the *legal* **responsibility** for the installation of vertical transportation. Therefore, we *must* review the latest codes available and *make sure* that all the requirements and regulations are carried through from the design stage to the construction stage of the project.

The following, in regard to code standards, is given for *information only.*

## Elevator Code Standards

Equipment provided under an elevator contract must be installed in a hoistway and machine room in accordance with code requirements applicable to the location of the installation. The following references from the **American Society of Mechanical Engineers Safety Code** for Elevators and Escalators, **ANSI/ASME** A17.1c-1986 **(an American National Standard),** and from the **Canadian Standards Association Standard,** Can3-B44-M85 Safety Code for Elevators **(a Canadian National Standard),** establish specific code compliance requirements that should be coordinated with the design criteria of the complete elevator installation in the United States and Canada, respectively.

## Hoistway Design

1.  *Enclosure.*  Hoistways must be enclosed for their entire height (except observation elevators) with fire-resistive enclosures such as masonry, concrete, or dry-wall construction. For multiple elevator installation, separate hoistways may be recommended or required. Pits shall be of noncombustible material designed to prevent entry of groundwater [Reference A17.1, Rules 100.1 and 100.2 (United States), and Reference B44, Sections 2.2.1 and 2.7.1 (Canada)].

2.  *Floor.*  A metal or concrete floor must be provided at the top of the hoistway [Reference A17.1, Rule 100.3 (United States), and Reference B44, Section 2.2.3 (Canada)].

3.  *Fire Protection.*  Hoistways of elevators shall be provided with means to prevent the accumulation of smoke and hot gases in case of fire as required by local or Model Building Codes [Reference A17-1, Rule 100.4 (United States), and Reference B44, Section 2.2.4 (Canada)]. Where air pressurization of the hoistway is utilized as a means of smoke and hot gas control, it shall be introduced into the hoistway without adversely affecting the elevator operation [Reference A17.1, Rule 100.4 (United States *only*)].

4.  *Walls.*  Hoistway walls shall not be provided with windows. Inside walls of enclosures shall have substantially flush surfaces without recesses or projections on sides not used for entrances to elevator. Setbacks or projections exceeding two (2) inches shall be beveled on the top at an angle of not less than seventy-five (75) degrees with the horizontal [Reference A17.1, Rule 100.6 (United States), and Reference B44, Section 2.2.5.1 (Canada)].

5.  ***Pipes or ducts.***   Pipes or ducts conveying gases, vapors, or liquids, and not used in connection with the operation of the elevator, shall not be installed in the hoistway, machine room, or machinery space [Reference A17.1, Rule 102.2 (United States), and Reference B44, Section 2.4.2 (Canada)].

## Machine Room Design

1.  ***Space.***   Spaces containing machines and control equipment shall be enclosed with fire-resistive enclosures and access doors which have a fire-resistance rating at least equal to that required for the hoistway enclosure [Reference A17.1, Rule 101.1 (United States), and Reference B44, Section 2.3.1 (Canada)].

2.  ***Equipment.***   Only machinery and equipment required for the operation of the elevator shall be permitted in the elevator machine room or space. For some installations, elevator equipment may be separated from other machinery or equipment by a grille enclosure with self-closing and self-locking door [Reference A17.1, Rules 101.2 and 300.2 (United States), and Reference B44, Section 2.3.2 (Canada)].

3.  ***Access.***   A permanent, safe, and convenient means of access shall be provided to the elevator machine room. Requirements depend upon design and location of machine room or space [Reference A17.1, Rule 101.3 (United States), and Reference B44, Section 2.3.3 (Canada)].

## Electrical Design

1.  ***Electrical equipment and wiring.***   All electrical equipment and wiring shall conform to Article 620 of the National Electrical Code, ANSI/NFFA 70-1987 (United States), and Article 38 of the Canadian Electrical Code Part I, CSA Standard C22.1-1986 (Canada).

2.  ***Main feeders.***   Main feeders for supplying power to the elevator shall be installed outside the hoistway. Only such electrical equipment used directly in connection with the elevator may be installed in the hoistway, machine room, or machinery space [Reference NEC Article 620-37 (United States), and CEC Section 38-014 (Canada)].

3.  ***Disconnects.***   A fused disconnect switch or circuit breaker, with feeder wiring, shall be provided for each elevator. The disconnect should be located in the vicinity of the Controller or Motor Generator Starter or on the lock jamb side of the machine room door (Canada) and be visible from the elevator machine [Reference NEC Article 620-51 (United States), and CEC 38-034 (Canada)].

4. *Lighting*

    *a.* Machine room shall be provided with permanent electric lighting and natural or mechanical ventilation [Reference A17.1, Rule 101.5 (United States), and Reference B44, Section 2.3.5 (Canada)].

    *b.* A permanent lighting fixture shall be provided in all pits, with a switch located to be accessible from the pit access door [Reference A17.1, Rule 106.1 (United States), and Reference B44, Section 2.7.5 (Canada)].

## V-4 VERTICAL TRANSPORTATION

Vertical transportation is a device for transporting people or goods from one floor to another floor. It applies to both of the following:

    *a.* **Vertical transportation within the structure.**

    *b.* **Vertical transportation outside the structure used to provide services within the structure.**

Vertical transportation may be grouped into three categories:

1. ***Elevators*** (V-5)

2. ***Vertical transportation system for handicapped persons*** (V-26)

3. ***Escalators and moving ramps*** (V-29 to V-36)

# ELEVATORS

## V-5   TYPES OF ELEVATORS

There are several types of elevators produced by elevator manufacturers in the United States and abroad.

In general, there are nine types of vertical transportation used in buildings:

1. ***Gearless traction elevators***
2. ***Geared traction elevators***
3. ***Hydraulic elevators***
4. ***Holeless hydraulic elevators***
5. ***Double-deck elevators***
6. ***Freight elevators***
7. ***Residential elevators***
8. ***Sidewalk elevators***
9. ***Dumbwaiters***

## V-6   GEARLESS TRACTION ELEVATORS (Fig. V-6)

Gearless traction elevators are commonly used in buildings ***over 10 stories*** in height.

The standard design for this type of elevator is as follows:

*a.*   **Car capacity** of 3000, 3500, and 4000 pounds.

*b.*   **Number of passengers per car** is 20, 23, and 27.

Number of passengers per car is based on 150 lb/passenger.

$$\frac{3000 \text{ lb.}}{20 \text{ pass.}} = 150 \text{ lb./passenger}$$

Very seldom can we find 20 persons in a 3000-lb car with an average weight of 150 lb.

*c.*   **The speed** in feet per minute (FPM) is 500, 700, 800, 1000, and 2000.

**The motion control** in this type of elevator s either DC direct drive or motor generator drive (a DC generator is used to produce direct current to operate the motor).

**The car finishes** are standard design, expanded design, or custom design (very expensive). They come with many optional features.

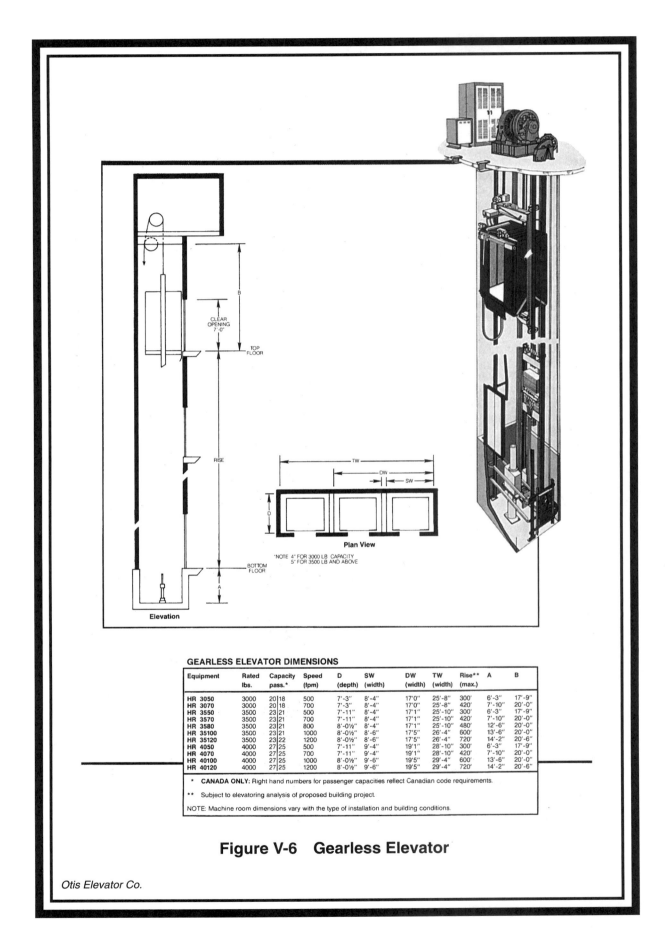

**GEARLESS ELEVATOR DIMENSIONS**

| Equipment | Rated lbs. | Capacity pass.* | Speed (fpm) | D (depth) | SW (width) | DW (width) | TW (width) | Rise** (max.) | A | B |
|---|---|---|---|---|---|---|---|---|---|---|
| HR 3050 | 3000 | 20\|18 | 500 | 7'-3" | 8'-4" | 17'0" | 25'-8" | 300' | 6'-3" | 17'-9" |
| HR 3070 | 3000 | 20\|18 | 700 | 7'-3" | 8'-4" | 17'0" | 25'-8" | 420' | 7'-10" | 20'-0" |
| HR 3550 | 3500 | 23\|21 | 500 | 7'-11" | 8'-4" | 17'1" | 25'-10" | 300' | 6'-3" | 17'-9" |
| HR 3570 | 3500 | 23\|21 | 700 | 7'-11" | 8'-4" | 17'1" | 25'-10" | 420' | 7'-10" | 20'-0" |
| HR 3580 | 3500 | 23\|21 | 800 | 8'-0½" | 8'-4" | 17'1" | 25'-10" | 480' | 12'-6" | 20'-0" |
| HR 35100 | 3500 | 23\|21 | 1000 | 8'-0½" | 8'-6" | 17'5" | 26'-4" | 600' | 13'-6" | 20'-0" |
| HR 35120 | 3500 | 23\|22 | 1200 | 8'-0½" | 8'-6" | 17'5" | 26'-4" | 720' | 14'-2" | 20'-6" |
| HR 4050 | 4000 | 27\|25 | 500 | 7'-11" | 9'-4" | 19'1" | 28'-10" | 300' | 6'-3" | 17'-9" |
| HR 4070 | 4000 | 27\|25 | 700 | 7'-11" | 9'-4" | 19'1" | 28'-10" | 420' | 7'-10" | 20'-0" |
| HR 40100 | 4000 | 27\|25 | 1000 | 8'-0½" | 9'-6" | 19'5" | 29'-4" | 600' | 13'-6" | 20'-0" |
| HR 40120 | 4000 | 27\|25 | 1200 | 8'-0½" | 9'-6" | 19'5" | 29'-4" | 720' | 14'-2" | 20'-6" |

\* **CANADA ONLY:** Right hand numbers for passenger capacities reflect Canadian code requirements.

\*\* Subject to elevatoring analysis of proposed building project.

NOTE: Machine room dimensions vary with the type of installation and building conditions.

**Figure V-6   Gearless Elevator**

*Otis Elevator Co.*

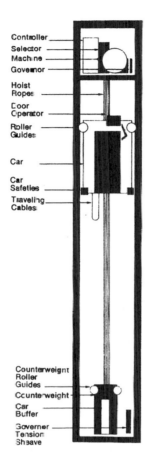

Controller
Selector
Machine
Governor
Hoist Ropes
Door Operator
Roller Guides
Car
Car Safeties
Traveling Cables
Counterweight Roller Guides
Counterweight
Car Buffer
Governer Tension Sheave

**Gearless Traction Elevators** are used in office buildings over 10 stories for speeds from 400 to 1,800 feet per minute. They require large, slow-speed electric motors (about 50 to 200 revolutions per minute), directly connected to a large, grooved drive sheave about 30 to 48 inches in diameter. One end of the "hoisting ropes" (actually steel cables), is attached to the top of the elevator, and then wrapped around the drive sheave in the special grooves. The other end of the hoisting ropes is attached to a counterweight that slides up and down in the shaftway on its own guide rails. With this weight, traction, and thus the lifting power is gained by the pressure of the wire ropes on the grooves of the sheave. Actually, the electric hoisting motor does not have to lift the full weight of the elevator car and its passengers. The weight of the car and about half its passenger load is balanced out by the counter-weight, which is sliding down as the car is going up. For speeds over 500 feet per minute, additional traction is gained by wrapping the hoisting ropes around a secondary sheave located just below the main drive sheave.

**Geared Traction Elevators** differ very little from the gearless type. The geared elevator is used usually at speeds from 25 to 350 feet per minute, and for loads up to 30,000 pounds or more. And, as the name implies, the system consists of a high-speed electric motor driving a worm and gear-type reduction unit, which drives the hoisting sheave. Such a design produces a relatively slow sheave speed with the high torque necessary for a wide range of passenger and freight elevator, and dumbwaiter applications. An electrically controlled brake between the motor and reduction unit stops the elevator to hold the car at floor level.

# Figure V-7  Components of Elevators

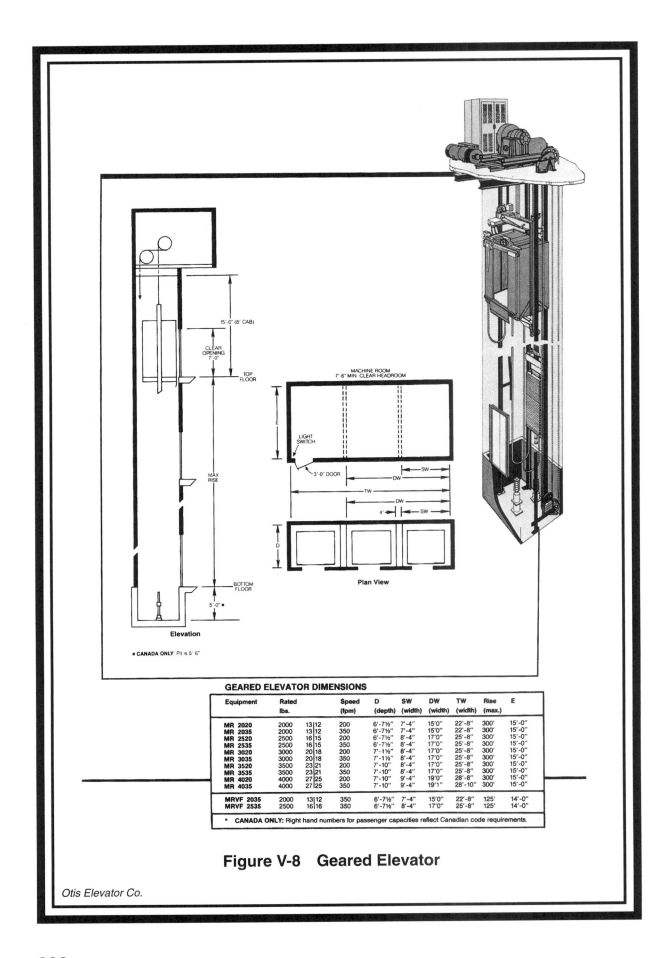

15'-0" (8' CAB)

CLEAR OPENING 7'-0"

TOP FLOOR

MAX RISE

BOTTOM FLOOR

5'-0" •

**Elevation**

• CANADA ONLY: Pit is 5' 6"

MACHINE ROOM 7'-6" MIN. CLEAR HEADROOM

LIGHT SWITCH

3'-0" DOOR

SW

DW

TW

DW

4'

SW

D

**Plan View**

**GEARED ELEVATOR DIMENSIONS**

| Equipment | Rated lbs. | | Speed (fpm) | D (depth) | SW (width) | DW (width) | TW (width) | Rise (max.) | E |
|---|---|---|---|---|---|---|---|---|---|
| MR 2020 | 2000 | 13\|12 | 200 | 6'-7½" | 7'-4" | 15'0" | 22'-8" | 300' | 15'-0" |
| MR 2035 | 2000 | 13\|12 | 350 | 6'-7½" | 7'-4" | 15'0" | 22'-8" | 300' | 15'-0" |
| MR 2520 | 2500 | 16\|15 | 200 | 6'-7½" | 8'-4" | 17'0" | 25'-8" | 300' | 15'-0" |
| MR 2535 | 2500 | 16\|15 | 350 | 6'-7½" | 8'-4" | 17'0" | 25'-8" | 300' | 15'-0" |
| MR 3020 | 3000 | 20\|18 | 200 | 7'-1½" | 8'-4" | 17'0" | 25'-8" | 300' | 15'-0" |
| MR 3035 | 3000 | 20\|18 | 350 | 7'-1½" | 8'-4" | 17'0" | 25'-8" | 300' | 15'-0" |
| MR 3520 | 3500 | 23\|21 | 200 | 7'-10" | 8'-4" | 17'0" | 25'-8" | 300' | 15'-0" |
| MR 3535 | 3500 | 23\|21 | 350 | 7'-10" | 8'-4" | 17'0" | 25'-8" | 300' | 15'-0" |
| MR 4020 | 4000 | 27\|25 | 200 | 7'-10" | 9'-4" | 19'0" | 28'-8" | 300' | 15'-0" |
| MR 4035 | 4000 | 27\|25 | 350 | 7'-10" | 9'-4" | 19'1" | 28'-10" | 300' | 15'-0" |
| MRVF 2035 | 2000 | 13\|12 | 350 | 6'-7½" | 7'-4" | 15'0" | 22'-8" | 125' | 14'-0" |
| MRVF 2535 | 2500 | 16\|16 | 350 | 6'-7½" | 8'-4" | 17'0" | 25'-8" | 125' | 14'-0" |

\* **CANADA ONLY:** Right hand numbers for passenger capacities reflect Canadian code requirements.

## Figure V-8   Geared Elevator

*Otis Elevator Co.*

332

## V-7 HOW DOES A GEARLESS TRACTION ELEVATOR WORK? (Fig. V-7)

They require large, slow-speed electric motors (about 50 to 200 revolutions per minute), directly connected to a large, grooved drive sheave about 30 to 48 inches in diameter.

One end of the **"hoisting ropes"** (actually steel cables) is attached to the top of the elevator and then wrapped around the drive sheave in the special grooves.

The other end of the hoisting ropes is attached to a counterweight that slides up and down in the shaftway on its own guide rails.

With this weight, traction—and thus the lifting power—is gained by the pressure of the wire ropes on the grooves of the sheave.

Actually, the electric hoisting motor does not have to lift the full weight of the elevator car and its passengers.

The weight of the car and about half its passenger load is balanced out by the counter-weight, which is sliding down as the car is going up.

For speeds over 500 feet per minute, additional traction is gained by wrapping the hoisting ropes around a secondary sheave located just below the main drive sheave.

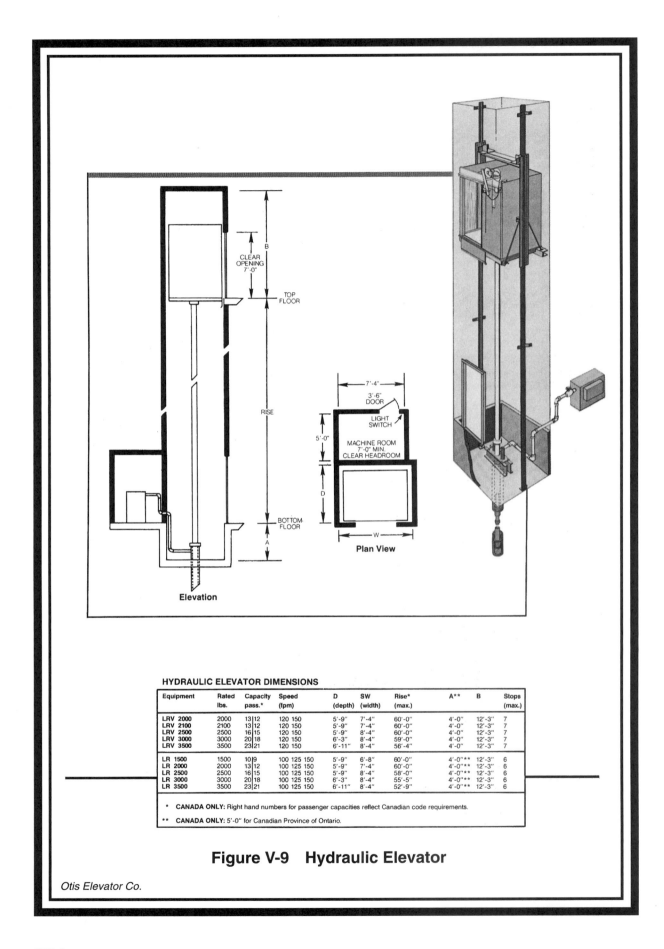

**HYDRAULIC ELEVATOR DIMENSIONS**

| Equipment | Rated lbs. | Capacity pass.* | Speed (fpm) | D (depth) | SW (width) | Rise* (max.) | A** | B | Stops (max.) |
|---|---|---|---|---|---|---|---|---|---|
| LRV 2000 | 2000 | 13\|12 | 120 150 | 5'-9" | 7'-4" | 60'-0" | 4'-0" | 12'-3" | 7 |
| LRV 2100 | 2100 | 13\|12 | 120 150 | 5'-9" | 7'-4" | 60'-0" | 4'-0" | 12'-3" | 7 |
| LRV 2500 | 2500 | 16\|15 | 120 150 | 5'-9" | 8'-4" | 60'-0" | 4'-0" | 12'-3" | 7 |
| LRV 3000 | 3000 | 20\|18 | 120 150 | 6'-3" | 8'-4" | 59'-0" | 4'-0" | 12'-3" | 7 |
| LRV 3500 | 3500 | 23\|21 | 120 150 | 6'-11" | 8'-4" | 56'-4" | 4'-0" | 12'-3" | 7 |
| LR 1500 | 1500 | 10\|9 | 100 125 150 | 5'-9" | 6'-8" | 60'-0" | 4'-0"** | 12'-3" | 6 |
| LR 2000 | 2000 | 13\|12 | 100 125 150 | 5'-9" | 7'-4" | 60'-0" | 4'-0"** | 12'-3" | 6 |
| LR 2500 | 2500 | 16\|15 | 100 125 150 | 5'-9" | 8'-4" | 58'-0" | 4'-0"** | 12'-3" | 6 |
| LR 3000 | 3000 | 20\|18 | 100 125 150 | 6'-3" | 8'-4" | 55'-5" | 4'-0"** | 12'-3" | 6 |
| LR 3500 | 3500 | 23\|21 | 100 125 150 | 6'-11" | 8'-4" | 52'-9" | 4'-0"** | 12'-3" | 6 |

\* **CANADA ONLY:** Right hand numbers for passenger capacities reflect Canadian code requirements.

\*\* **CANADA ONLY:** 5'-0" for Canadian Province of Ontario.

## Figure V-9  Hydraulic Elevator

Otis Elevator Co.

334

**V-8    GEARED TRACTION ELEVATORS (Fig. V-8)**

Geared traction elevators are generally used in the buildings which don't require high-speed service, such as apartment buildings *8 to 20 stories* in height, with a speed of 200 to 300 FPM. They also can be used for a rise of 300 ft or 25 stories.

The standard design for this type of elevator is as follows:

a.    **Car capacity** of 2000, 2500, 3000, 3500, and 4000 pounds.

b.    **The number of passengers per car** is 13, 16, 20, and 27.

c.    **The speed** in feet per minute (FPM) is 350, 400, 450, and 500.


**V-9    HOW DOES A GEARED TRACTION ELEVATOR WORK? (Fig. V-7)**

The geared traction elevator's operation is very similar to the gearless traction type.

The geared elevator can be used for loads from 30 to 30,000 pounds and perhaps more.

The speed for heavy loads is 25 to 350 FPM.

The system consists of a big-speed electric motor driving a worm- and gear-type reduction unit, which drives the hoisting sheave. Such a design produces a relatively slow sheave speed with the high torque necessary for a wide range of passenger and freight elevators and dumbwaiter applications.

An electrically controlled brake between the motor and reduction unit stops the elevator to hold the car at floor level.


**V-10    HYDRAULIC ELEVATORS (Fig. V-9)**

Hydraulic elevators are used exclusively in low-rise buildings *up to 5 stories* in height.

They do not require any penthouse or overhead support for machinery.

The standard design for this type of elevator is as follows:

a.    **A car capacity** of 1500, 2000, 2100, 2500, 3000, and 3500 pounds.

b.    **The number of passengers per car** is 10, 13, 16, 20, and 23.

c.    **The speed** in feet per minute (FPM) is 100 and 150.

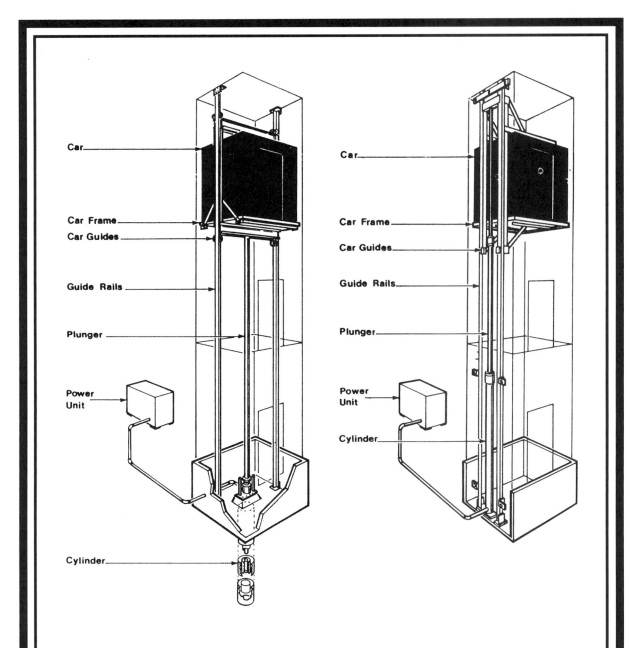

**Hydraulic Elevators** are used extensively today in low-rise buildings, like those up to five stories high. With speeds rarely exceeding 150 feet per minute, the big advantage of the hydraulic elevators is that they do not need any penthouse or overhead support for machinery. The elevator is mounted on a direct plunger or piston, in a cylinder that extends into the ground to a depth equal to the height that the elevator will rise. Of relatively simple operation, the system requires only an electric pump to force the oil into the cylinder to raise the elevator. Electrically controlled valves release the oil from the cylinder to give the elevator a controlled descent.

### Figure V-10   Hydraulic Elevators

## V-11  HOW DOES A HYDRAULIC ELEVATOR WORK? (Fig. V-10)

The elevator is mounted on a direct plunger or piston in a cylinder that extends into the ground to a depth equal to the height that the elevator will rise.

Of relatively simple operation, the system requires only an electric pump to force the oil into the cylinder to raise the elevator.

Electrically controlled valves release the oil from the cylinder to give the elevator a controlled descent.

## V-12  HOLELESS HYDRAULIC ELEVATORS (Fig. V-11)

Holeless hydraulic elevators work similarly to hydraulic elevators (Fig. V-10).

They are used when the ground under the location of the elevator consists of solid hard rock and drilling for a direct plunger or piston becomes very expensive, and also at the existence of subsoil water or unstable sand which may cause lateral movement of the plunger.

The standard design for this type of elevator is as follows:

a.   **A car capacity** of 2000, 2100, and 2500 pounds.

b.   **The number of passengers per car** is 13 and 16.

c.   **The speed** in feet per minute (FPM) is 100.

This type of elevator is used for *up to 2 stories.*

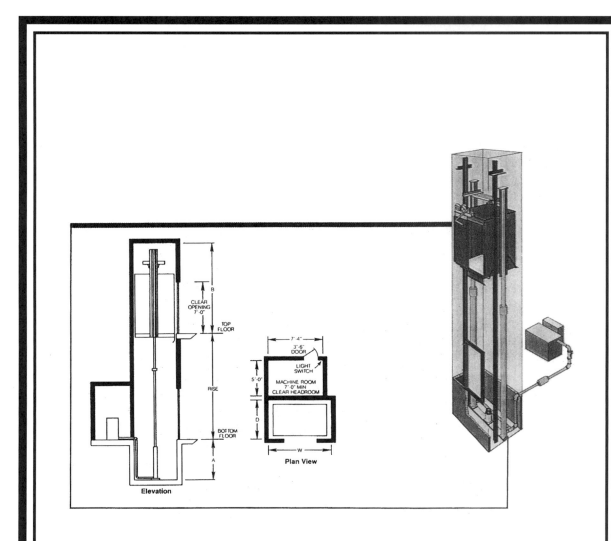

Elevation

Plan View

**HOLELESS ELEVATOR DIMENSIONS**

| Equipment | Rated lbs. | Capacity pass.* | Speed (fpm) | D (depth) | W (width) | Rise (max.) | A● | B | Stops (max.) |
|-----------|-----------|-----------------|-------------|-----------|-----------|-------------|-----|-----|--------------|
| LRV 2010L | 2000 | 13│12 | 100 | 5'-9" | 7'-4" | 15'-0" | 4'-0"● | 12'-4" | 2 |
| LRV 2110L | 2100 | 13│12 | 100 | 5'-9" | 7'-4" | 15'-0" | 4'-0"● | 12'-4" | 2 |
| LRV 2510L | 2500 | 16│15 | 100 | 5'-9" | 8'-4" | 15'-0" | 4'-0"● | 12'-4" | 2 |

●  Pit depth (A) increases 1" for each 1" increase in rise over 13'-8" to a maximum pit depth of 5'-4" for the maximum rise of 15'-0" except starting pit depth for Canadian Province of Ontario is 5'-0".

*  CANADA ONLY: Right hand numbers for passenger capacities reflect Canadian code requirements.

# Figure V-11   Holeless Elevator

## V-13    DOUBLE-DECK ELEVATORS (Fig. V-12)

Double-deck elevators consist of two elevator cars mounted one on top of the other, and they move at the same time.

People entering the lobby of the building will get on the lower car if they want to go to an odd-numbered floor in the building. Those who want to go to an even-numbered floor will take an escalator to the mezzanine level in the lobby to get on the upper car.

People coming down can take either car to the level they desire.

**This type of elevator system will save valuable space and improve elevator service.**

In a 40-story building, 18 double-deck elevators are equivalent to 26 commonly used elevators, with comparable elevator handling capacity. At the same time, it will add approximately 27,000 square feet of rentable space to the building.

The Citicorp Tower in New York City, which was completed in 1977, uses 20 double-deck elevators and two regular elevators to handle passenger service.


## V-14    FREIGHT ELEVATORS (Fig. V-13)

Freight elevators are used to move furniture, building maintenance materials, goods, etc., in commercial buildings.

They are also used for motor vehicle and truck loading in garages and industrial buildings.

For the maximum **rise of 60 feet,** a *hydraulic elevator* may be used.

**They are available** in 3000-, 4000-, 6000-, 8000-, 10,000-, and 12,000-lb duties and a speed of 50 feet per minute.

For tall buildings, *geared traction elevators* are used with a **capacity** of 2500- to 12,000-lb duties and **speeds** between 50 and 200 feet per minute.

*Looking Up the bank of five glass double-deck observation elevators installed in the new government office building on the Sparks Street Mall in Ottawa, Canada. Units operate in a glass atrium allowing passengers to see through the glass-faced building. It is the first such application of double-deck observation elevators in the world.*

## Figure V-12   Double-Deck Elevators

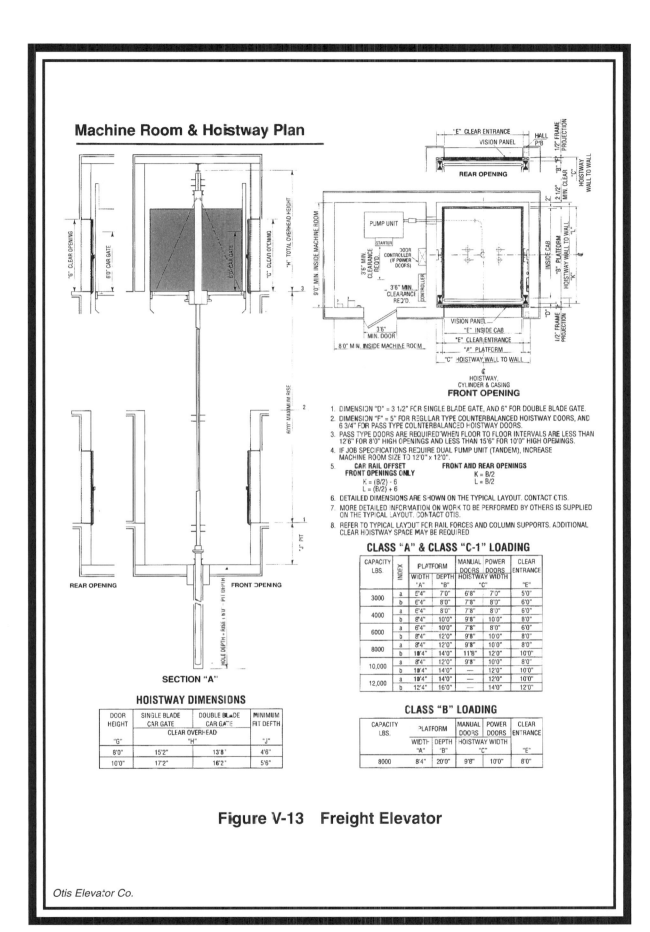

# Machine Room & Hoistway Plan

**SECTION "A"**

**REAR OPENING**     **FRONT OPENING**

**REAR OPENING**

**FRONT OPENING**

1. DIMENSION "D" = 3 1/2" FOR SINGLE BLADE GATE, AND 6" FOR DOUBLE BLADE GATE.
2. DIMENSION "F" = 5" FOR REGULAR TYPE COUNTERBALANCED HOISTWAY DOORS, AND 6 3/4" FOR PASS TYPE COUNTERBALANCED HOISTWAY DOORS.
3. PASS TYPE DOORS ARE REQUIRED WHEN FLOOR TO FLOOR INTERVALS ARE LESS THAN 12'6" FOR 8'0" HIGH OPENINGS AND LESS THAN 15'6" FOR 10'0" HIGH OPENINGS.
4. IF JOB SPECIFICATIONS REQUIRE DUAL PUMP UNIT (TANDEM), INCREASE MACHINE ROOM SIZE TO 12'0" x 12'0".
5. 

| CAR RAIL OFFSET | |
|---|---|
| **FRONT OPENINGS ONLY** | **FRONT AND REAR OPENINGS** |
| K = (B/2) - 6 | K = B/2 |
| L = (B/2) + 6 | L = B/2 |

6. DETAILED DIMENSIONS ARE SHOWN ON THE TYPICAL LAYOUT. CONTACT OTIS.
7. MORE DETAILED INFORMATION ON WORK TO BE PERFORMED BY OTHERS IS SUPPLIED ON THE TYPICAL LAYOUT. CONTACT OTIS.
8. REFER TO TYPICAL LAYOUT FOR RAIL FORCES AND COLUMN SUPPORTS. ADDITIONAL CLEAR HOISTWAY SPACE MAY BE REQUIRED

## CLASS "A" & CLASS "C-1" LOADING

| CAPACITY LBS. | INDEX | PLATFORM | | MANUAL DOORS | POWER DOORS | CLEAR ENTRANCE |
|---|---|---|---|---|---|---|
| | | WIDTH "A" | DEPTH "B" | HOISTWAY WIDTH "C" | | "E" |
| 3000 | a | 6'4" | 7'0" | 6'8" | 7'0" | 5'0" |
| | b | 6'4" | 8'0" | 7'8" | 8'0" | 6'0" |
| 4000 | a | 6'4" | 8'0" | 7'8" | 8'0" | 6'0" |
| | b | 8'4" | 10'0" | 9'8" | 10'0" | 8'0" |
| 6000 | a | 6'4" | 10'0" | 7'8" | 8'0" | 6'0" |
| | b | 8'4" | 12'0" | 9'8" | 10'0" | 8'0" |
| 8000 | a | 8'4" | 12'0" | 9'8" | 10'0" | 8'0" |
| | b | 10'4" | 14'0" | 11'8" | 12'0" | 10'0" |
| 10,000 | a | 8'4" | 12'0" | 9'8" | 10'0" | 8'0" |
| | b | 10'4" | 14'0" | — | 12'0" | 10'0" |
| 12,000 | a | 10'4" | 14'0" | — | 12'0" | 10'0" |
| | b | 12'4" | 16'0" | — | 14'0" | 12'0" |

## CLASS "B" LOADING

| CAPACITY LBS. | PLATFORM | | MANUAL DOORS | POWER DOORS | CLEAR ENTRANCE |
|---|---|---|---|---|---|
| | WIDTH "A" | DEPTH "B" | HOISTWAY WIDTH "C" | | "E" |
| 8000 | 8'4" | 20'0" | 9'8" | 10'0" | 8'0" |

## HOISTWAY DIMENSIONS

| DOOR HEIGHT | SINGLE BLADE CAR GATE | DOUBLE BLADE CAR GATE | MINIMUM PIT DEPTH |
|---|---|---|---|
| | CLEAR OVERHEAD | | |
| "G" | "H" | | "J" |
| 8'0" | 15'2" | 13'8" | 4'6" |
| 10'0" | 17'2" | 16'2" | 5'6" |

# Figure V-13   Freight Elevator

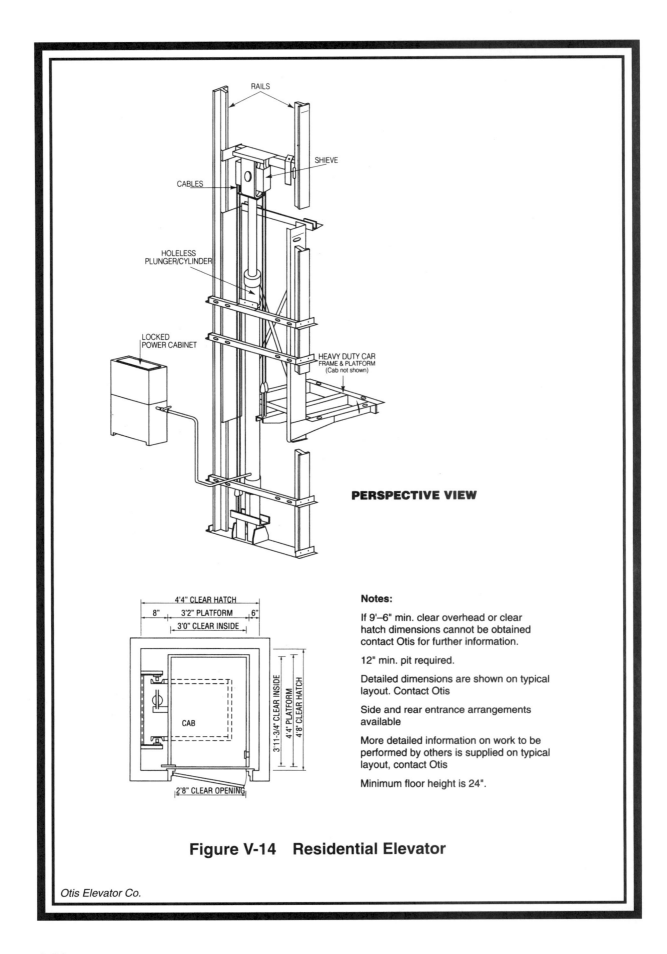

RAILS

SHIEVE

CABLES

HOLELESS
PLUNGER/CYLINDER

LOCKED
POWER CABINET

HEAVY DUTY CAR
FRAME & PLATFORM
(Cab not shown)

**PERSPECTIVE VIEW**

4'4" CLEAR HATCH

8"  3'2" PLATFORM  6"

3'0" CLEAR INSIDE

CAB

3'11-3/4" CLEAR INSIDE

4'4" PLATFORM

4'8" CLEAR HATCH

2'8" CLEAR OPENING

**Notes:**

If 9'–6" min. clear overhead or clear
hatch dimensions cannot be obtained
contact Otis for further information.

12" min. pit required.

Detailed dimensions are shown on typical
layout. Contact Otis

Side and rear entrance arrangements
available

More detailed information on work to be
performed by others is supplied on typical
layout, contact Otis

Minimum floor height is 24".

# Figure V-14   Residential Elevator

The **ANSI** code for elevators has established three **load classifications** for freight elevators:

Class A    ***Loading by hand truck.***   The rated load is 50 lb/ft$^2$ of net inside the platform area. A single item may not exceed the car's rated load.

Class B    ***Motor vehicle loading.***   The rated load is 50 lb/ft$^2$.

Class C    ***Industrial truck loading.***   The rated load is 50 lb/ft$^2$.

## V-15    RESIDENTIAL ELEVATORS (Fig. V-14)

Residential elevators require approximately 20 square feet of space.

They use a two-step hydraulic power unit.

They use single-phase 120/220 V.

They can be installed in standard wood-frame or block walls.

Cars are available in standard design or they can be designed by the architect.

## V-16    SIDEWALK ELEVATORS (Fig. V-15)

Sidewalk elevators are used mainly in an areaway outside the building for delivery of goods and material between the sidewalk and basement(s).

**The capacity** of the car is 500 to 5000 lb at a speed of 20 feet per minute.

**The usual car size** is 4 ft × 4 ft and requires a 4-ft pit.

**It operates** by a turn key and hold switch or push button and hold switch from the sidewalk.

SIDEWALK

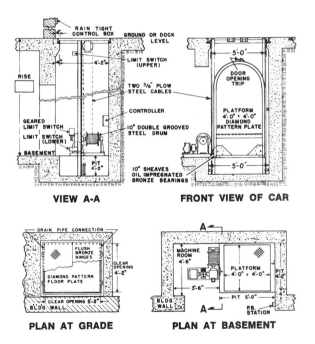

**TYPE PSH SIDEWALK** Car size usually up to 7′ x 5′ and capacity up to 10,000#. Larger sizes available.

**TYPE PFH FREIGHT** Car size & capacity to suit any requirement 10′-0″ x 20′-0″ and more than 10,000# ideal with (4) guide rails and (2) hydraulic cylinders. Synchronized & directly coupled to car structure.

VIEW A-A

FRONT VIEW OF CAR

PLAN AT GRADE

PLAN AT BASEMENT

# Figure V-15   Sidewalk Elevator

**344**

**DUMBWAITERS (Fig. V-16)**

Dumbwaiters are also known as *ejection lifts.*

They are used for vertical movement of food carts, dishes, books, linens, garbage,* etc.

They are installed in buildings which require constant movement of goods and materials from one floor to another such as restaurants, hospitals, libraries, etc.

* For vertical movement of garbage and waste materials, a completely separate dumbwaiter is required.

There are two types of dumbwaiter machines:

1.    **Traction-type**
2.    **Drum-type**

**The capacity** is 50, 100, 200, 300, 500 lb, up to a maximum of 1000 lb, with a two- or three-side opening, and **speeds** of 40 to 300 feet per minute.

It requires a single-phase 120/240 V.

A 500-lb dumbwaiter has a **platform width** of 36 in., depth of 36 in., and a height of 48 in.

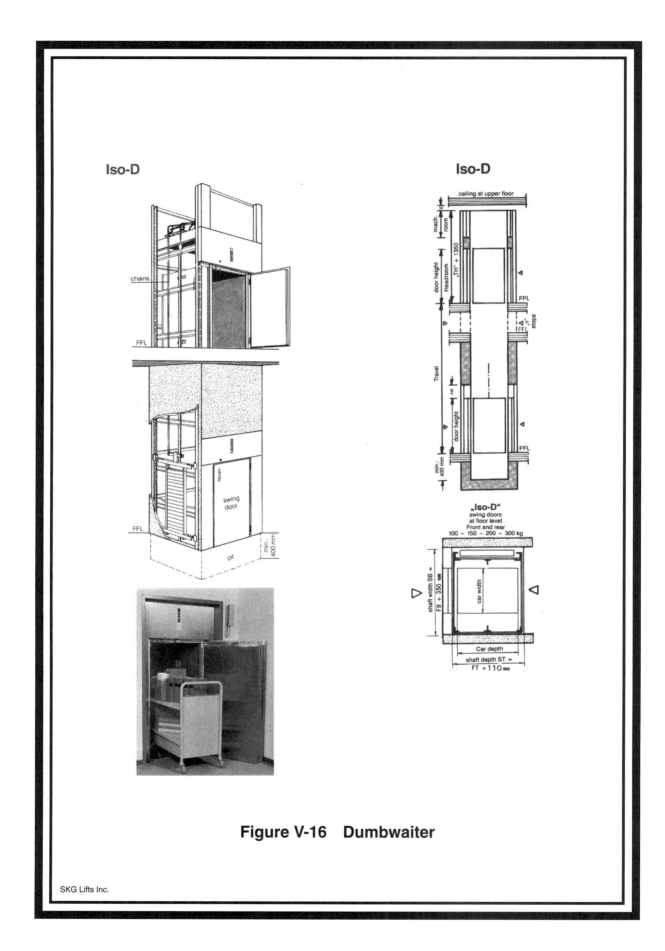

**Figure V-16  Dumbwaiter**

SKG Lifts Inc.

# ELEVATOR SELECTION

### V-18   OBJECTIVES FOR SELECTING ELEVATORS

The selection of elevators for any type of multistory building requires the following:

1.  ***Quantity of elevators for proper service***

    a.  Minimum waiting period

    b.  Loading and unloading time

    c.  Number of elevators in system

    d.  Carrying capacity of each car

    e.  Speed of the elevators

2.  ***Quality of elevators***

    a.  Interior design of the cab

    b.  Proper and adequate lighting

    c.  Communication

    d.  Visual floor indicator

    e.  Comfortable acceleration and smooth deceleration

3.  ***Economics***

    Approximate cost of elevator system in a 20- to 60-story building is 10.8 to 12.4 percent of total construction cost.

    a.  **Overdesign will increase the cost.**

    b.  **Underdesign is not acceptable.**

The criteria for the proper selection of elevator systems are:

1.  ***Interval*** (I)
2.  ***Handling capacity*** (HC)
3.  ***Round-trip time*** (RT)

| Building Type | APPROXIMATE FOR ELEVATORS | |
|---|---|---|
| | Intervals (I)<br>per second | Handling Capacity (HC)<br>% of population to be<br>moved in 5 minutes |
| **APARTMENT** | | |
| Luxury | 50 | 7 |
| Middle-income | 70 | 8 |
| Low-income | 80 | 8 |
| **HOTEL** | | |
| Luxury | 50 | 12 |
| Average | 60 | 10 |
| **HOSPITAL** | 50 | 17 |
| **OFFICE BUILDING** | | |
| Prestige | 26 | 13 |
| General | 35 | 17 |

**Figure V-17   Intervals and Handling Capacity**

**V-19    INTERVAL (*I*)**

Interval in seconds is the period of time a passenger has to wait for the arrival of the elevator on any floor of the building.

The approximate interval in seconds which is generally acceptable is given in Fig. V-17.

**V-20    HANDLING CAPACITY (HC)**

Handling capacity is the percentage of total population of the building to be moved by the elevator system (Fig. V-17).

It is based on the number of persons moved in 5 minutes or 300 seconds.

**Handling capacity is determined by two factors:**

1.    **Car size (number of passengers per car)**

2.    **Interval**

**For the *system\* which has *more than one* elevator:**

$$HC = \frac{300 \text{ sec} \times \text{no. of passenger per car}}{I}$$

or          $$HC = \frac{300\ P}{(I)} \qquad (1)$$

For the *system\* which has *only one* elevator, the interval (*I*) is equal to the round-trip (RT):

$$I = RT$$

Therefore

$$h = \frac{300\ P}{RT} \qquad (2)$$

\* The word ***system*** is used to express the "elevator system" in a building, whether it has one or more elevators.

**HC** = 5 minutes (300 seconds) **handling capacity,** used for a system with **more than one elevator**

*h* = 5 minutes (300 seconds) handling capacity, used for a system with **only one elevator**

**P** = Number of persons in car

**RT** = Round-trip time in seconds

*I* = Interval in seconds

**349**

| APARTMENT | No. of persons per bedroom |
|---|---|
| Luxury | 1.4 |
| Middle-income | 2.0 |
| Low-income | 3.0 |

| HOTEL | No. of persons per room |
|---|---|
| Luxury | 1.4 |
| Average | 1.9 |

| HOSPITAL | No. of persons per bed |
|---|---|
| Private | 1.5 |
| Public | 3.5 |

| OFFICE BUILDING | Square foot per person |
|---|---|
| Prestige | 130 |
| Typical | 120 |

**Figure V-18   Average Population Density**

**AT** = Average trip time in seconds

**N** = Number of cars in system

**D** = Population density in square feet per person

## V-21 ROUND-TRIP TIME (RT)

Round-trip time consists of four factors:

1. Running time in seconds at rated speed

2. Passenger loading and unloading at each stop

3. Time to open and close the door(s) in seconds

4. Time to accelerate and decelerate in seconds

## V-22 NUMBER OF CARS IN SYSTEM

In a system which has more than one car, the interval ($I$) will be reduced in direct proportion to the number of the cars in the system.

$$\text{Therefore} \qquad I = \frac{RT}{N} \qquad (3)$$

For the system which has more than one car, the handling capacity (HC) is equal to the handling capacity of the system with only one car ($h$) times the number of cars ($N$) in the system.

$$HC = N \times h$$

$$\text{then} \qquad N = \frac{HC}{h} \qquad (4)$$

|  | Car Capacity<br>Pounds | Car Speed<br>FPM | No. of Persons<br>per car |
|---|---|---|---|
| **GEARED ELEVATOR** | | | |
| | 2000 | 350 | 12 |
| | 2500 | 350-400-450-500 | 15 |
| | 3000 | 350-400-450-500 | 18 |
| | 3500 | 350-400-450-500 | 21 |
| | 4000 | 350 | 25 |
| **GEARLESS ELEVATOR** | | | |
| | 3000 | 500-700 | 18 |
| | 3500 | 500-700-800-1000-1200 | 21 |
| | 4000 | 500-700-1000-1200 | 25 |

**Figure V-19   Elevators' Car Capacity and Speed**

| | Car Capacity Pounds | | Height of Rise Feet | Car Speed FPM |
|---|---|---|---|---|
| APARTMENT* | 2500 | | 125 | 350 |
| | 3000 | | 175 | 400 |
| | | | 250 | 500 |
| | | | 350 | 700 |
| HOTEL | 2500 | | 125 | 350 |
| | 3000 | | 175 | 500 |
| | | | 250 | 700 |
| | | | 350 | 800 |
| HOSPITAL | 4000 | | 125 | 350 |
| | | | 175 | 400 |
| | | | 350 | 700 |
| OFFICE BUILDING | 2500 | | 125 | 350 |
| | 3000 | | 175 | 500 |
| | | | 250 | 700 |
| | 3500 | | 350 | 800 |
| | | above | 350 | 1000 - 1200 |

* FHA minimum 2 elevator per building and 120 bedrooms per car.

Figure V-20   Suggested Elevator Selection

### V-23 TRAVEL TIME (T T)

The average travel time is approximately

$$\text{Travel time} = \tfrac{1}{2}\ \text{interval} + \text{travel time to desired floor}$$

In residential, hotel, recreational buildings, etc., people are relaxed; therefore, they may tolerate a longer travel time. Therefore, intervals up to 80 seconds are acceptable.

However, in a commercial building people are busy and excited; therefore, they are not able to tolerate a long travel time. For this reason, the intervals should not exceed 35 seconds. Please see V-17.

In general travel time:

less than 60 seconds is **desirable.**

up to 75 seconds is **acceptable.**

up to 90 seconds is **annoying.**

of 120 seconds is the **limit of toleration.**

### V-24 DESIGN CRITERIA FOR SELECTING ELEVATORS

To select an elevator system for any type of building we need to know the following:

1. **Number of floors above lobby**

2. **Net square feet of each floor**

3. **Floor-to-floor height**

4. **Type and usage**

**Step 1. Find Interval ($I$)**

by using Fig. V-17. The figures given may be adjusted by plus or minus 5 percent.

**Step 2. Find the average population density (APD).**

a.  If you have already determined the total population for your building, use your figure and don't follow steps 2 and 3.

b.  If you know APD for your building, use your figure.

c.  If you don't have item a or b, then you may use Fig. V-18 to obtain APD for your building.

**Step 3. Find the total population of the building:**

**Total net area ft²/F × no. of F = total net area of building**

**Total net area of building ÷ APD ft²/P = population of building**

where   F = floor

**Step 4. Find the handling capacity (HC).**

Use Fig. V-17 to find percent of HC for your building.

**HC = population of building × % of HC**

**Step 5. Find the height that elevators are to serve.**

**(No. of F) × (F-to-F height) = height or rise**

**Step 6. Select a car size.**

*a.* Figure V-19 gives the car capacity in pounds, car speed in FPM, and **number of persons per car (P)** for geared and gearless elevators.

*b.* Figure V-20 gives the suggested elevator selection.

*c.* Use the rise of your building found in step 5 to guide you in choosing the car capacity and speed required.

This selection will be checked in step 10.

**Step 7. Find round-trip (RT).**

Use Fig. V-21.

*a.* Use speed of car found in step 6.

*b.* Use the number of floors the system is serving (given).

*c.* Use the floor-to-floor height and capacity of the car (given) and **determine the RT in seconds.**

**Step 8. Find the handling capacity of the system (h) assuming it has only one car.**

Using Eq. (2):

$$h = \frac{300\ P}{RT}$$

We found the value (P) in step 6 and RT in step 7.

**Step 9. Find the number of cars (*N*) in the system**

by using Eq. (4).

$$N = \frac{HC}{h}$$

**Step 10. Check the interval (*I*).**

Using Eq. (3):

$$I = \frac{RT}{N}$$

*a.*     If the interval is 5 percent above or below interval found in Step 1, **you have chosen the right-size car(s) at the right speed and an adequate number of cars for the system.** *Stop* **and go to step 11.**

*b.*     **If the interval is below 5 percent, you have overdesigned the system.**
To correct this:

    1.  Reduce the number of cars *or*

    2.  Use smaller-size cars *or*

    3.  Reduce the speed.

*c.*     **If the interval is above 5 percent, you have underdesigned the system.**
To correct this:

    1.  Add to the number of cars *or*

    2.  Use larger-size cars *or*

    3.  Choose higher speed, if possible.

**Step 11. A.** Use Fig. V-6 for gearless or Fig. V-8 for geared elevator to obtain dimensions. For depth and width of the car(s) and elevator(s) shaft(s) or

    1.  Choose the elevator manufacturer you like to consider.

    2.  Use their latest catalog.

    3.  Normally they are willing to assist you without charge with the understanding that you will allow them to participate in the bidding phase.

    **B.** You have to use your judgment in regard to the arrangement of the elevators on your floor plan. The lobby between elevators may be 8, 9, or 10 ft, depending on the population it is serving.

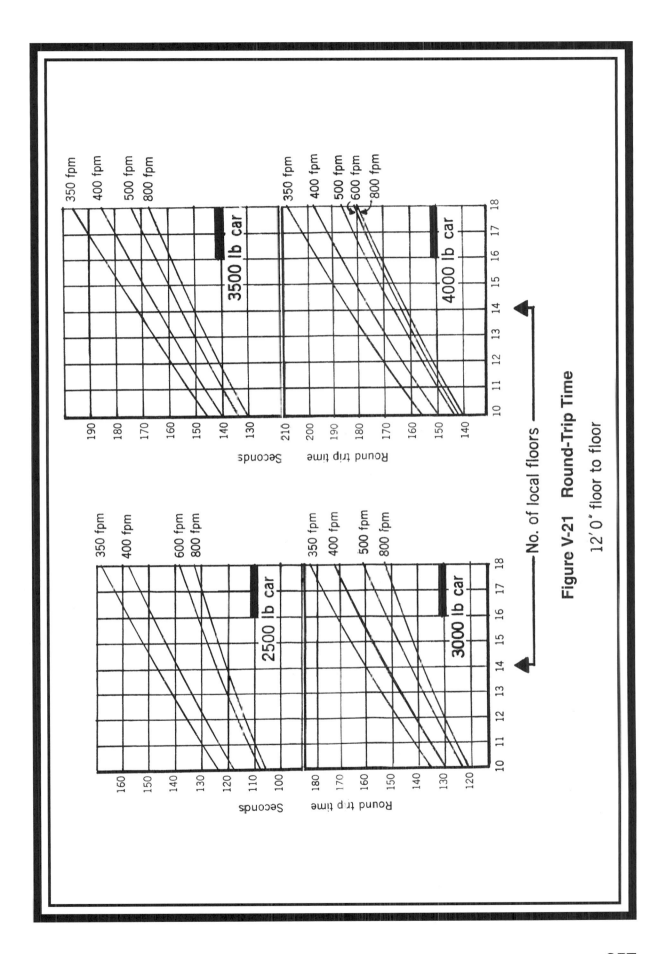

**Figure V-21   Round-Trip Time**

12' 0" floor to floor

## V-25 EXAMPLE FOR SELECTION OF ELEVATORS

### Problem

Select an elevator system for a 17-story prestige office building.

Total rentable area consists of 16 stories (F) above the lobby, and the net area of each floor is 10,000 ft$^2$. The floor-to-floor height is 12 ft.

### Solution

**Step 1. Find Interval (*I*).**

Using Fig. V-17:

$$I = \textbf{26 seconds}$$

**Step 2. Find average population density (APD).**

Using Fig. V-18:

$$\textbf{APD} = \textbf{120 ft}^2\textbf{/person}$$

**Step 3. Find total population of building.**

$$10,000 \text{ ft}^2/\text{F} \times 16 \text{ F} = 160,000 \text{ ft}^2$$

Population = 160,000 ft$^2$ ÷ 120 ft$^2$/person = **1333 persons**

**Step 4. Find handling capacity (HC).**

Using Fig. V-17:

$$HC = 13\%$$

$$HC = 1333 \text{ persons} \times 0.13 = \textbf{173 persons}$$

**Step 5. Find the height which the elevators are to serve.**

$$16 \text{ F} \times 12 \text{ ft/F} = \textbf{192 ft height}$$

**Step 6.** Select a car size.

Using Figs. V-19 and V-20:

Use 3000-lb car capacity

700 FPM car speed

18 persons (P) per car

**Step 7. Find round-trip (RT).**

Using Fig. V-21:

For 3000-lb car, 12 ft floor to floor

16 local floors and 700 FPM

RT = 155 seconds

**Step 8.** Find handling capacity of the system assuming it has only one car.

Using Eq. (2): $\quad h = \dfrac{300P}{RT} \qquad h = \dfrac{300(18)}{155} = 34.83$

**Step 9. Find number of cars in system**

Using Eq. (4): $\quad N = \dfrac{HC}{h} \qquad N = \dfrac{173}{34.83} = 4.96$ cars

**Step 10. Check the interval ($I$)**

Using Eq. (3): $\quad I = \dfrac{RT}{N} \qquad I = \dfrac{155}{5} = 31$ seconds

Interval required, step 1, $I = 26$ seconds. Under design, try 6 cars.

$$I = \frac{155}{6} = 25.83 \text{ seconds } \textbf{OK}$$

Use 6 – 3000-lb cars, 700 FPM speed, gearless elevators.

Place 3 cars side by side opposite 3 cars side by side with a 10 ft. lobby between the row of elevators. The size of each shaft is:

**width = TW = 25′-8″**

**depth = D = 7′-3″**

The total area required on each floor for this elevator system is:

**depth = 7′-3″ + 7′-3″ + 10′ = 24″-6″**

**width = 25′-8″.**

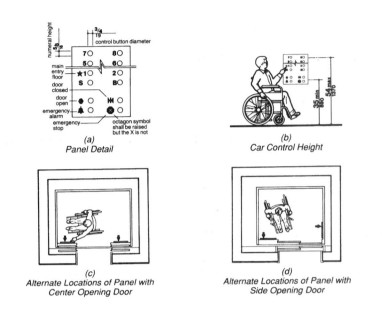

*(a)*
*Panel Detail*

*(b)*
*Car Control Height*

*(c)*
*Alternate Locations of Panel with*
*Center Opening Door*

*(d)*
*Alternate Locations of Panel with*
*Side Opening Door*

**Figure 23 as shown in the ADAAG**
**Car Controls**

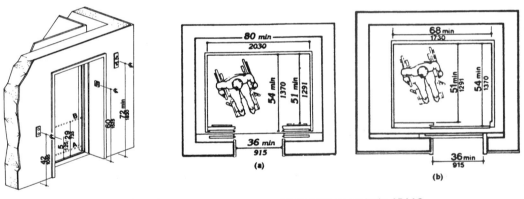

NOTE: The automatic door reopening device is
activated if an object passes through either line A
or line B. Line A and Line B represent the vertical
locations of the door reopening device not requiring contact.

**Figure 20 as shown in the ADAAG**
**Hoistway and Elevator Entrances**

*(a)*

*(b)*

**Figure 22 as shown in the ADAAG**
**Minimum Dimensions of Elevator Cars**

# Figure V-22    Minimum Dimensions of Elevator Cars
## for Handicapped Persons

American National Standards Institute, Inc.

# VERTICAL TRANSPORTATION SYSTEMS
# FOR HANDICAPPED PERSONS

**V-26    AMERICANS WITH DISABILITIES ACT (ADA)**

This Federal law was enacted in July of 1990. Its purpose is to ensure greater accessibility to

a.    **Public entities**

b.    **Places of public accommodation**

c.    **Commercial facilities, etc.**

for approximately 43 million disabled Americans. The **ADA** includes:

a.    **New construction**

b.    **Certain types of alterations or renovations**

c.    **Existing structures**

**Public entities.**   Examples include state and local government facilities, airports, and transportation facilities.

**Places of public accommodation.**   Examples include hotels, restaurants, theaters, service establishments (doctors' and lawyers' offices, shopping centers, etc.), public transportation terminals, parks, zoos, and other places of recreation, schools, day care centers, health spas, and hospitals.

**Commercial facilities.**   Examples include factories, warehouses, office buildings, or any facilities where employment may occur.

In general, any building where business is conducted or where a service is rendered for the benefit of the general public will be required to comply to various degrees.

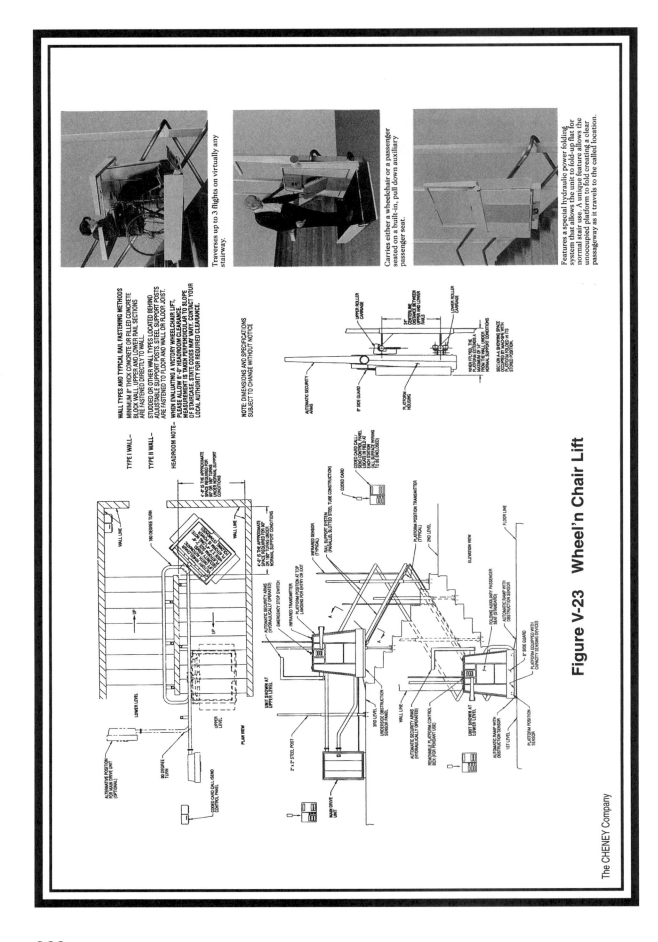

Traverses up to 3 flights on virtually any stairway.

Carries either a wheelchair or a passenger seated on a built-in, pull down auxiliary passenger seat.

Features a special hydraulic power folding system that allows the unit to fold-up flat for normal stair use. A unique feature allows the unoccupied platform to fold creating a clear passageway as it travels to the called location.

**Figure V-23   Wheel'n Chair Lift**

**Exemptions**

For places of public accommodation and commercial facilities, elevators are not required in facilities of less than three stories or with less than 3000 square feet per story unless the facility is:

a.   A shopping center or shopping mall, or the professional office of a health care provider.

b.   A terminal, depot, or other station used for specified public transportation, or an airport passenger terminal.

There are other exemptions and interpretations allowed under the guidelines. The best way to find out how they affect your specific situation is to consult your legal advisors.

**Technical Standards** (Fig. V-22)

Public accommodations and commercial facilities (as defined in the ADA) are subject to ADAAG (Americans with Disabilities Act Accessibility Guidelines).

Certain public entities may be subject to either ADAAG or UFAS (Uniform Federal Accessibility Standard), depending on the specific situation.

State and local accessibility laws, which may add additional requirements, also have to be taken into account.

**V-27   WHEELCHAIR LIFT (Fig. V-23)**

Wheelchair lifts are used to carry a person in a wheelchair, or a seated passenger, from one floor to another floor by the stairway.

They move on custom-built parallel steel tubes fastened to concrete walls or other structurally sound materials.

They should meet the requirements set forth by ANSI Sec. A-17.1.

The capacity is up to 500 pounds, and they must have an overload-sensing device.

The speed is generally 25 feet per minute on straight sections, and 12 feet per minute on curved sections.

The platforms are approximately 32 × 42 in. with a nonskid platform (Fig. V-18).

## V-28    VERTICAL WHEELCHAIR LIFT (Fig. V-24)

A vertical wheelchair lift is an easy stairway access for handicapped persons.

They can be used indoors or outdoors, and they must meet the ANSI Sec. A-17.1 requirements.

The capacity is 500 pounds with a speed of 12 feet per minute.

The traveling distance is 12 feet, and platforms are approximately 12 $ft^2$.

Power requirements are 6 amp at 240 V, single-phase.

# Installation Drawing

## Handi-Enclosure Installation Drawing

NOTES:
1) ENCLOSURE IS NOT INTENDED FOR USE WHERE FIRE RATED CONSTRUCTION IS REQUIRED.
2) IT IS ESSENTIAL THAT FLOOR BELOW ENCLOSURE AND RAMP IS LEVEL.
3) REQUIRED LIGHTING ABOVE ENCLOSURE—OTHERS—
4) FOR PITTED APPLICATIONS CONTACT AUTHORIZED CHENEY DISTRIBUTOR FOR DETAILS.

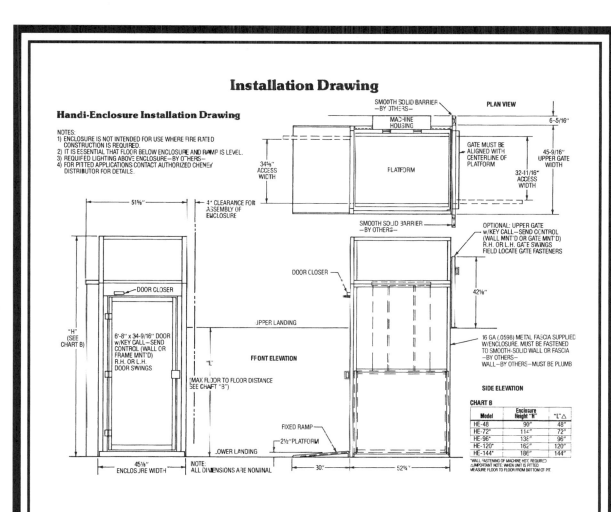

**PLAN VIEW**

SMOOTH SOLID BARRIER —BY OTHERS—

MACHINE HOUSING

PLATFORM

6-5/16"

GATE MUST BE ALIGNED WITH CENTERLINE OF PLATFORM

45-9/16" UPPER GATE WIDTH

32-11/16" ACCESS WIDTH

34⅝" ACCESS WIDTH

SMOOTH SOLID BARRIER —BY OTHERS—

51⅜"

4" CLEARANCE FOR ASSEMBLY OF ENCLOSURE

OPTIONAL: UPPER GATE w/KEY CALL—SEND CONTROL (WALL MNT'D OR GATE MNT'D) R.H. OR L.H. GATE SWINGS FIELD LOCATE GATE FASTENERS

DOOR CLOSER

DOOR CLOSER

42⅛"

UPPER LANDING

"H" (SEE CHART B)

6'-8" x 34-9/16" DOOR w/KEY CALL—SEND CONTROL (WALL OR FRAME MNT'D) R.H. OR L.H. DOOR SWINGS

**FRONT ELEVATION**

"L"

(MAX FLOOR TO FLOOR DISTANCE SEE CHART "B")

16 GA (.0598) METAL FASCIA SUPPLIED W/ENCLOSURE. MUST BE FASTENED TO SMOOTH-SOLID WALL OR FASCIA —BY OTHERS— WALL—BY OTHERS—MUST BE PLUMB

**SIDE ELEVATION**

**CHART B**

| Model | Enclosure Height "H" | "L" △ |
|---|---|---|
| HE-48 | 90" | 48" |
| HE-72* | 114" | 72" |
| HE-96* | 138" | 96" |
| HE-120* | 162" | 120" |
| HE-144* | 186" | 144" |

*WALL FASTENING OF MACHINE HS'G REQUIRED
△IMPORTANT NOTE: WHEN UNIT IS PITTED MEASURE FLOOR TO FLOOR FROM BOTTOM OF PIT.

FIXED RAMP

2½" PLATFORM

LOWER LANDING

45⅛" ENCLOSURE WIDTH

NOTE: ALL DIMENSIONS ARE NOMINAL

30"

52⅝"

# Figure V-24   Vertical Wheelchair Lift

# ESCALATORS AND MOVING RAMPS

**V-29    ESCALATORS**

Escalators are also called *moving electric stairways.*

The escalator was operated for the first time at the Paris Exposition in 1900. It was designed by the Otis Elevator Company.

Wherever large volumes of people must be moved quickly and freely from one floor to another, floor escalators are the solution in vertical transportation.

They are commonly used in office buildings, hotels, shopping centers, department stores, etc.

**V-30    CODE REQUIREMENTS**

The following are given as guidelines; you **must** check state and local code requirements.

Dimensional data and safety devices **must** comply with ANSI/ASME A 17.1 and, in Canada, with 3-B44-M85.

State and local code requirements **must** be used if they are more stringent.

**ANSI/ASME A 17.1**

*Section 800—Protection of Floor Openings*

Rule 800.1    **Protection Required**

Floor openings for escalators shall be protected against the passage of flame, heat, and/or smoke in accordance with the provisions of the building code.

*Section 801.1—Protection of Trusses and Machine Spaces Against Fire*

Rule 801.1    **Protection Required**

The sides and undersides of escalator trusses and machinery spaces shall be enclosed in fire-resistive materials. Means may be provided for adequate ventilation of the driving machine and control spaces.

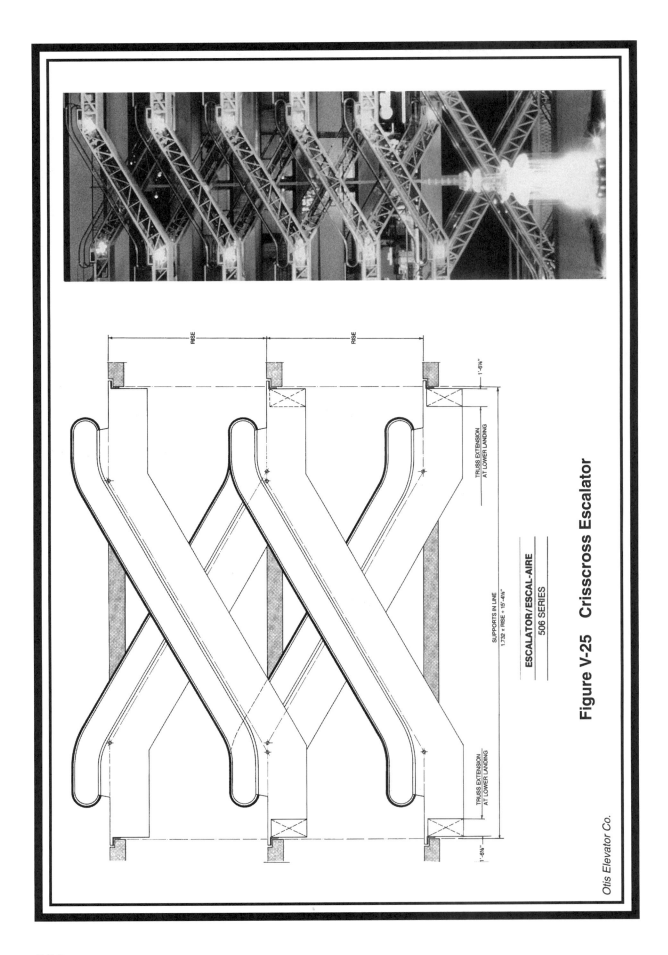

ESCALATOR/ESCAL-AIRE
506 SERIES

RISE

RISE

1'-6¼"

TRUSS EXTENSION
AT LOWER LANDING

SUPPORTS IN LINE
1.732 × RISE + 15'-4¼"

TRUSS EXTENSION
AT LOWER LANDING

1'-6¼"

**Figure V-25   Crisscross Escalator**

*Section 806—Lighting Access and Electrical Work*

Rule 806.1   **Lighting of Machine Room**

Rule 806.1b   **Machine in Truss**

A permanent 20-A grounding-type 110-V duplex receptacle accessibly located within the machine area of the truss shall be provided to accommodate a drop-cord light.

Rule 806.2   **Lighting of Step Treads**

Step treads shall be illuminated throughout their run. The light intensity on the treads shall be in accordance with local codes and ordinances for stairways.

**CAN 3-B44-M85**

*Section 8—Escalators*

8.1   **Protection of Floor Openings**

Floor openings for escalators shall be protected against the passage of flame, smoke, or gasses in the event of fire, in accordance with local ordinances. Where there are no building codes, local by-laws, or ordinances, the National Building Code of Canada shall apply.

8.2   **Protection of Trusses and Machinery Spaces Against Fires**

8.2.1   The sides and undersides of escalators, trusses, and machinery spaces shall be enclosed in fire-resistive materials.

8.2.2   Means may be provided for adequate ventilation of the driving machine and control spaces.

8.7   **Lighting and Access**

8.7.1   **Lighting of Machine Room**

8.7.1.1   Permanent artificial lighting shall be provided in every machine room.

8.7.1.2   The lighting switch shall be so located that it can be operated without passing over or reaching over any part of the machinery.

8.7.2   **Lighting of Step Treads**

8.7.2.1   Step treads shall be illuminated throughout their run.

8.7.2.2   The light intensity on the tread surfaces shall be not less than 20 lx.

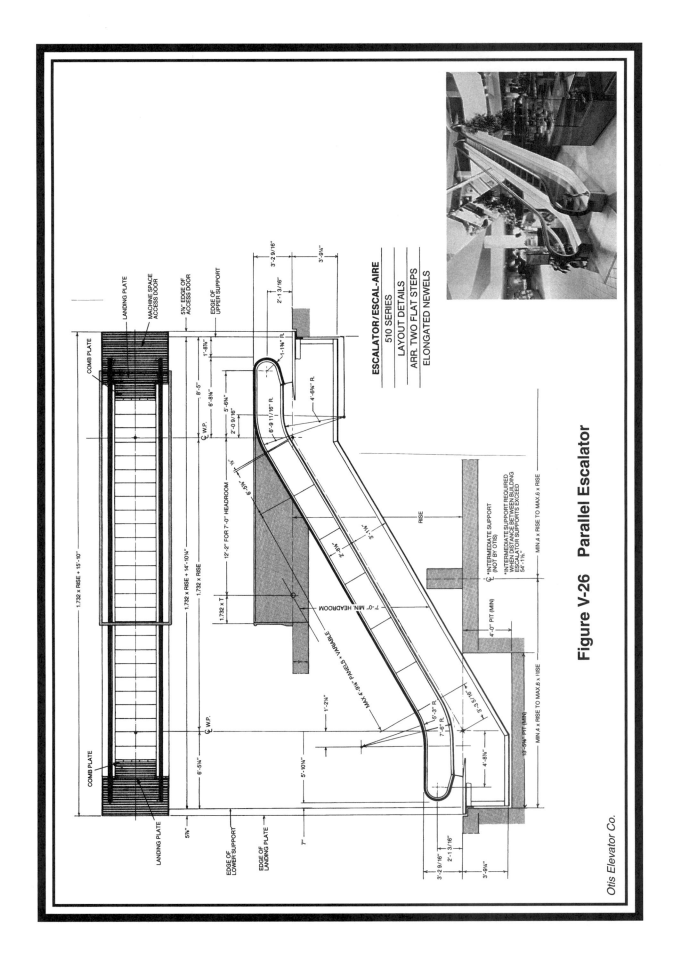

**ESCALATOR/ESCAL-AIRE**
510 SERIES
LAYOUT DETAILS
ARR. TWO FLAT STEPS
ELONGATED NEWELS

## Figure V-26  Parallel Escalator

*Otis Elevator Co.*

## V-31    ESCALATOR ARRANGEMENTS

Escalators can be arranged in two different ways in a building:

1.    *Crisscross escalators*

2.    *Parallel escalators*


## V-32    CRISSCROSS ESCALATORS (Fig. V-25)

In this arrangement, the escalators going up and down crisscross each other, allowing the riders a continuous ride without mixing with floor traffic.

Crisscross escalators' advantages, when compared with parallel escalators:

a.    **They are lower in overall cost.**

b.    **They require less space.**

c.    **They have lower structural requirements.**

They are used commonly to serve more than two floors.


## V-33    PARALLEL ESCALATORS (Fig. V-26)

In this arrangement, the escalators going up and down are parallel to each other. If used to serve more than two floors, the riders have to walk 180° or the entire length of the escalator in order to continue with their ride.

Parallel escalators' advantages are:

a.    **They have an impressive appearance.**

b.    **When several escalators are used, they can serve the traffic both up and down at peak periods by reversing their movements.**

They are used commonly in banks, transportation terminals, etc.

| | | | | |
|---|---|---|---|---|
| Type of machine: | AC induction motor with worm gear reducer in truss. | | | |
| Size: | 32 | 40 | 48 | |
| Step width: | 24″ | 32″ | 40″ | |
| Speed: | 100 feet per minute | | | |
| Maximum rise: | 20′-0″ | | | |
| Capacity (Pass/hr): | 2,250 | 3,375 | 4,500 | |

For higher rise applications, choose the Otis 510. It accommodates rises up to 32′-10″ at a 100 feet per minute operating speed.

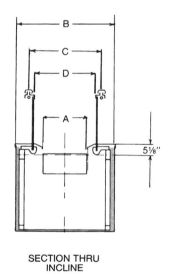

SECTION THRU
INCLINE

**WIDTHS**

| SIZE | A | B | C | D |
|---|---|---|---|---|
| 32 | 2′-0″ | 3′-11½″ | 3′-0¼″ | 2′-8″ |
| 40 | 2′-8″ | 4′-7½″ | 3′-8¼″ | 3′-4″ |
| 48 | 3′-4″ | 5′-3½″ | 4′-4¼″ | 4′-0″ |

**ESCALATOR/ESCAL-AIRE**
**506 SERIES**

**WIDTHS**

| SIZE | A | B | C | D |
|---|---|---|---|---|
| 32 | 2′-0″ | 4′-2⅜″ | 2′-10½″ | 2′-6⅞″ |
| 40 | 2′-8″ | 4′-10⅜″ | 3′-6½″ | 3′-2⅞″ |
| 48 | 3′-4″ | 5′-6⅜″ | 4′-2½″ | 3′-10⅞″ |

**ESCALATOR/ESCAL-AIRE**
**510 SERIES**

## Figure V-27   Escalators' Size, Speed, and Dimensions

*Otis Elevator Co.*

**CAPACITY, SIZE, AND SPEED OF ESCALATORS**

**Capacity:** The capacity of an escalator in passengers per hour is dependent on the speed that it travels and the width of the step. A realistic step loading to be used in planning is one passenger per step on a 40-inch-wide step and ½ passenger on a 24-inch-wide step.

At 100 FPM, the size 48 escalator with a 40-inch-wide step can transport 4500 passengers/hour. The size 32 escalator with a 24-inch-wide step carries 2250 passengers/hour.

Although more than one passenger can stand on a 40-inch-wide step, psychological factors, reaction times, and personal habits all discourage this type of loading on a moving escalator.

Only under unique and controlled conditions should capacity deviate from the recommendations.

**Width:** The width of an escalator is the width of the step tread in inches. This is a recent change in the ANSI/ASME A 17.1 code.

**The old width designations for 32 or 48 are now referred to as *size,* and have corresponding step widths of 24 and 40 inches, respectively.**

**Rises:** Each escalator model has rise limitations that are the result of design or performance conditions. To exceed these limits requires engineering and manufacturing efforts that will increase the cost and lead time. In some cases, performance may also be below standard, reducing product life and increasing maintenance.

Please see Fig. V-27 for size, step width, speed, maximum rise, and capacity (passengers/hr).

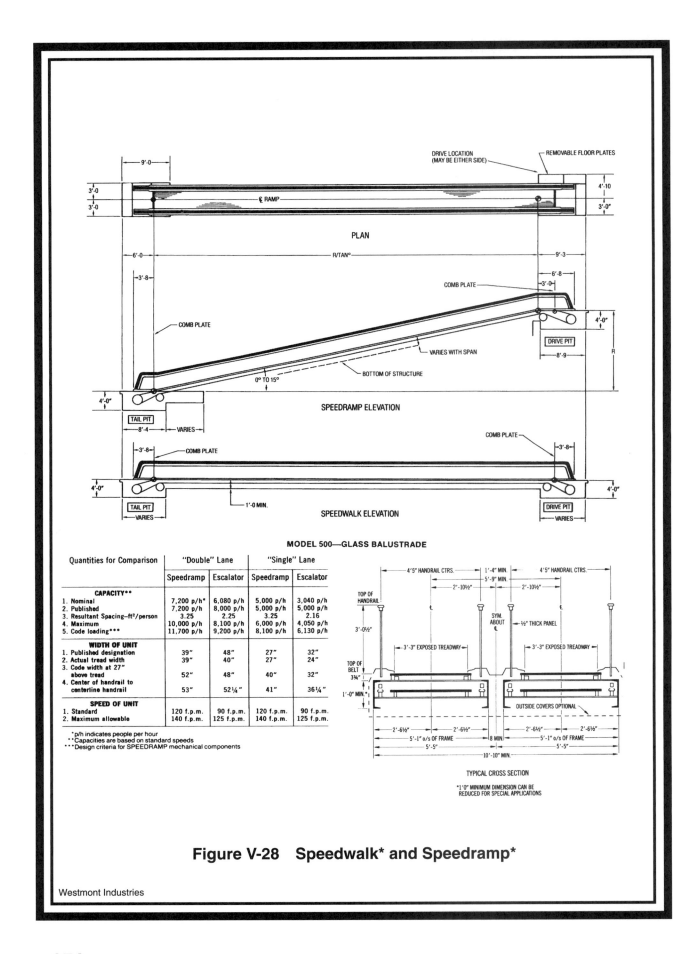

PLAN

SPEEDRAMP ELEVATION

SPEEDWALK ELEVATION

**MODEL 500—GLASS BALUSTRADE**

| Quantities for Comparison | "Double" Lane | | "Single" Lane | |
|---|---|---|---|---|
| | Speedramp | Escalator | Speedramp | Escalator |
| **CAPACITY**** | | | | |
| 1. Nominal | 7,200 p/h* | 6,080 p/h | 5,000 p/h | 3,040 p/h |
| 2. Published | 7,200 p/h | 8,000 p/h | 5,000 p/h | 5,000 p/h |
| 3. Resultant Spacing—ft²/person | 3.25 | 2.25 | 3.25 | 2.16 |
| 4. Maximum | 10,000 p/h | 8,100 p/h | 6,000 p/h | 4,050 p/h |
| 5. Code loading*** | 11,700 p/h | 9,200 p/h | 8,100 p/h | 6,130 p/h |
| **WIDTH OF UNIT** | | | | |
| 1. Published designation | 39" | 48" | 27" | 32" |
| 2. Actual tread width | 39" | 40" | 27" | 24" |
| 3. Code width at 27" above tread | 52" | 48" | 40" | 32" |
| 4. Center of handrail to centerline handrail | 53" | 52¼" | 41" | 36¼" |
| **SPEED OF UNIT** | | | | |
| 1. Standard | 120 f.p.m. | 90 f.p.m. | 120 f.p.m. | 90 f.p.m. |
| 2. Maximum allowable | 140 f.p.m. | 125 f.p.m. | 140 f.p.m. | 125 f.p.m. |

*p/h indicates people per hour
**Capacities are based on standard speeds
***Design criteria for SPEEDRAMP mechanical components

TYPICAL CROSS SECTION

"1'0" MINIMUM DIMENSION CAN BE
REDUCED FOR SPECIAL APPLICATIONS

# Figure V-28  Speedwalk* and Speedramp*

## V-35    DESIGN CRITERIA FOR SELECTION OF ESCALATORS

**Step 1.** Determine the maximum number of persons who have to be moved from floor to floor in 1 hour.

**Step 2.** Use Fig. V-27 to determine the number of escalator(s) and size.

> Provide adequate landing space at the top and bottom of the escalator.

### Example

In an office building 8600 persons have to be moved from the lobby level to the mezzanine level in one hour. What is the size and number of escalators required?

### Solution

Using Fig. V-27:

> Two 48-in. escalators may be used, good for 9000 persons per hour.
>
> However, if we use three 40-in. escalators, we would create more flexibility.

> *a.*    All three escalators can operate to move people up and down during the peak hours.
>
> *b.*    Two escalators may be run in one direction and one escalator in the opposite direction when needed
>
> *c.*    Two escalators may operate, one up and one down, and the third one can be shut off when traffic is slow.


## V-36    MOVING RAMPS AND WALK (Fig. V-28)

Moving ramp and walks are used to move people in a controlled and continuous direction between two points.

Their widths are 27 and 32 in. for single lane or 39 and 48 in. for double lane.

The moving ramp may be used with a maximum 15° incline angle.

The moving walk may be used for up to a 10° incline angle.

The speed limit is between 90 and 140 FPM with a capacity of between 3000 to 7000 persons per hour.

# List of Tables

**378**

# List of Charts and Diagrams

# Index

Absorptance, 35
AC generator, 107
Accessible, 35
Alive, 35
Alternating current, 9, 107, 112
Alternator, 35
Ambient temperature, 35
American National Standard Institute (ANSI), 325
American Wire Gauge (AWG), 35, 159
Americans with Disability Act (ADA), 361
Ammeter, 35
Ampacity, 35
Ampère, André-Marie, 6, 15
Ampere (A), 15, 35
Ampère's law, 15
Amplification, 35
Amplifier, 35
Amplitude, 36
Anode, 36
Antenna, 36
Apothecary weights, 18
Appliance, 36
Appliance outlet, 36
Approved, 36
Arago, D. F., 6
Arc, 36
Arc lighting, 13
Arc resistance, 36
Arcing time, 36
Armature, 36
Armor, 36

Askarel, 36
Automatic, 37
Automatic transfer equipment, 37
Autotransformer, 37
Auxiliary, 37
Avoirdupois weights, 18

Ballast, 37
Bar, 37
Bare, 37
Base, 37
Base load, 37
Battery, 37, 124
    dry-cell, 124
    Edison cell, 125
    solar, 125
    storage, 124
Becquerel, A. E., 235
Belt, 37
Belted-type cable, 37
Boiler, 5
Branch circuit, 38
    appliance, 38
    general purpose, 38
    individual, 38
Breaker strip, 38
Breakout, 38
British thermal unit (Btu), 38, 120
Brush, C. F., 7, 235
Brush, 38
Buna, 38

**381**

ASKA

NORTH
AMERICA

CALIFORNIA

P A C I F I C

O C E A N

QUEBEC
CHICAGO • BOSTON

LONDON
EUROPE

A T L A N T I C

LOUISIANA
FLORIDA

YUCATAN
GUATEMALA      JAMAICA
HONDURAS

AFRICA

PERU

O C E A N

SOUTH
AMERICA

3:00  4:00  5:00  6:00  7:00  8:00  9:00  10:00  11:00  12:00  1:00  2:00

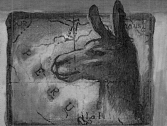

# SOMEWHERE IN THE
# WORLD RIGHT NOW

## by Stacey Schuett

Alfred A. Knopf  New York

*To Mavis and Frances,*
*for seeing beyond the horizon*

THIS IS A BORZOI BOOK PUBLISHED BY ALFRED A. KNOPF, INC.

Copyright © 1995 by Stacey Schuett
All rights reserved under International and Pan-American Copyright Conventions.
Published in the United States of America by Alfred A. Knopf, Inc., New York,
and simultaneously in Canada by Random House of Canada Limited, Toronto.
Distributed by Random House, Inc., New York.

Maps from *Goode's World Atlas* copyright © 1994 by Rand McNally, R.L. 91-S-244.
Used by permission of Rand McNally & Company

Library of Congress Cataloging-in-Publication Data

Schuett, Stacey.
Somewhere in the world right now / written and illustrated by Stacey Schuett.
p.   cm.
Summary: Describes what is happening in different places around the
world at a particular time.
ISBN 0-679-86537-3 (trade hardcover)—ISBN 0-679-96537-8 (lib. bdg.)
[1. Time—Fiction. 2. Geography—Fiction.] I. Title.
PZ7.S3855So   1994
[E]—dc20   94-6223

Manufactured in the United States of America
10 9 8 7 6 5 4 3 2 1

# A Note to the Reader

The earth is constantly turning on its axis. One full rotation takes 24 hours, and during that time, the position of the sun in the sky changes, creating light and dark, day and night.

It used to be that many places measured time from when the sun was highest overhead, at noon. But this meant that neighboring cities would have different clock times at the same instant. Faster travel and better communications made this inconvenient, so in 1884 an international agreement was reached to adopt a system of standard times. The planet was divided into 24 equal areas called time zones. (However, as the maps on the endpapers show, the time zones were adjusted to conform to geographical and political boundaries.) Everywhere within each zone, it is always the same time. When moving from one time zone to the next, there is a change in one hour, forward or backward, depending on whether one is moving east or west.

An important feature of the time zone system is the international date line. As the sun crosses the 24 time zones, somewhere a change in date must occur. The international community needed to agree on a boundary where the new date would begin. It's clear that to have the date change in the middle of a crowded country would cause much confusion, so the line was drawn through the Pacific Ocean, with a few jogs around inhabited regions. Since the earth turns eastward, places west of the date line move into the new date first. So, no matter where in the world you are, there's somewhere else where it's a different day!

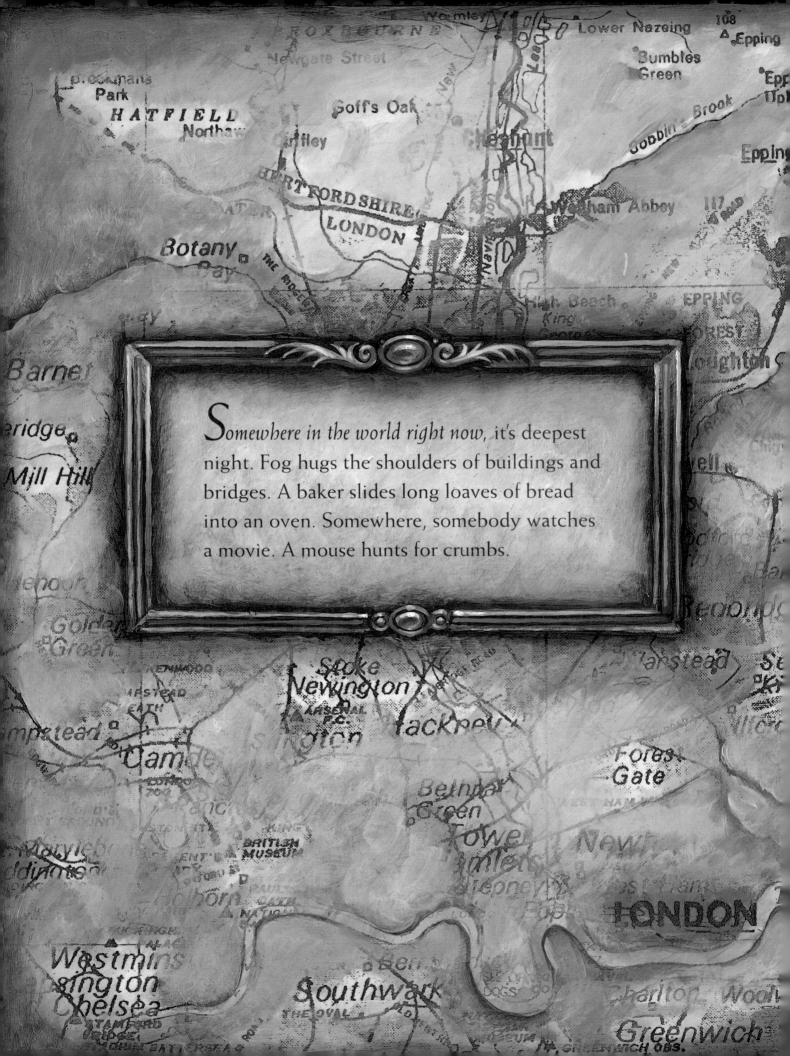

*Somewhere in the world right now,* it's deepest night. Fog hugs the shoulders of buildings and bridges. A baker slides long loaves of bread into an oven. Somewhere, somebody watches a movie. A mouse hunts for crumbs.

Somewhere else, in velvety darkness, elephants sleep standing up, swaying gently from side to side.

Whales breach and dive in the sea, singing their low,
sweet songs. Penguins press close to their chicks to
keep them warm.

Somewhere, the night wind sighs and murmurs.
The moon shines through a window. A little girl
lies dreaming of tomorrow.

But somewhere else, right now, tomorrow is already here. Dawn is breaking. A rooster crows and people are waking up.

Somewhere, it's morning. Families eat breakfast. Kids get dressed for school. Farmers leave for the fields to tend their crops.

Birds stretch their wings and sing. A boy hides
a note for a friend to find. People go to work,
stores open . . . another day has begun.

And somewhere, right now, people buy food for their
midday meal. A carpenter measures wood. A dog runs
off with his lunch.

In the late morning shade, a baby kangaroo
naps. A koala munches eucalyptus leaves.

Somewhere in the world right now, fishing boats return with their catch. Sea gulls swoop and dive and bicker over scraps.

Somewhere, the afternoon sun gilds workers in the vineyards. Cows graze and rest in the grass. A girl tends her horses.

Somewhere in the world right now, the sun is setting. Monkeys screech from the trees as a jaguar glides through the jungle. Parrots mutter and chuckle. Shadows grow long.

North Chicago
Lake Forest
Highland Park
Winnetka
Wilmette
Evanston
CHICAGO
E.
Chicago

And somewhere else, the sun has already slipped away, drawing evening down like a shade. In the city, signs flash on and off, off and on. Trains whoosh through tunnels, taking people home.

Watseka

Powler

Somewhere, suddenly, houses light up. A girl and her brother race each other to the door.

The day's work is done. A family sits down to supper.

LOUISIANA

Lake
Charles

Biloxi

And somewhere else, night has fallen. Clouds
cushion the moon as it climbs across the sky.
An alligator sleeps in an inky swamp beneath
a tapestry of stars.

The last notes of a song fade into the rhythm of the waves.

Somewhere, a truck driver follows her headlights
down a lonely road. A night watchman starts his shift.

Friends say good night.

Somewhere, somebody reads a story, and someone listens.

Brookline

Voices whisper, "Sweet dreams," and under a blanket of night, lights go out one by one, somewhere in the world . . .

right now.